# Modern Livestock Production

# Modern Livestock Production

Misha Peralta

R CALLISTO REFERENCE

www.callistoreference.com

**Callisto Reference,**
118-35 Queens Blvd., Suite 400,
Forest Hills, NY 11375, USA

Visit us on the World Wide Web at:
www.callistoreference.com

ISBN: 978-1-64116-535-8 (Hardback)

**Cataloging-in-Publication Data**

Modern livestock production / Misha Peralta.
 p. cm.
Includes bibliographical references and index.
ISBN 978-1-64116-535-8
1. Livestock. 2. Animal culture. 3. Domestic animals. 4. Animal industry. I. Peralta, Misha.
SF61 .M63 2022
636--dc23

# Table of Contents

Permissions

Index

# Preface

This book has been written, keeping in view that students want more practical information. Thus, my aim has been to make it as comprehensive as possible for the readers. I would like to extend my thanks to my family and co-workers for their knowledge, support and encouragement all along.

Livestock are the domesticated animals that are raised in an agricultural setting for the production of labor and commodities such as milk, meat, fur, leather and wool. The branch of agriculture which deals with the raising, selective breeding and maintenance of livestock is termed as animal husbandry. It is necessary to maintain animal health to maximize the production of livestock and the products derived from them. Livestock management also includes the control, treatment and prevention of diseases which can be caught by livestock such as foot and mouth disease, classical swine fever and scrapie. The farming practices in livestock management are broadly classified as intensive or extensive. Intensive livestock farming involves housing animals in high density conditions, focusing on maximizing output while minimizing costs. Animals are left to roam at will or under the supervision of a herdsman under extensive livestock farming. This book explores all the important aspects of livestock production and management. It elucidates new techniques and their applications in a multidisciplinary approach. This book will serve as a valuable source of reference for those interested in this field.

A brief description of the chapters is provided below for further understanding:

Chapter – Livestock Production

The domesticated animals that are raised in an agricultural setting to produce labor and commodities such as meat, milk, eggs, fur, etc. are known as livestock. Some of the common types of livestock production system are grassland-based livestock production system and landless ruminant production system. This chapter has been carefully written to provide an easy understanding of these livestock production systems.

Chapter – Types of Livestock

Some of the diverse types of livestock are cattle, domestic sheep, domestic pigs, goats, donkeys, camels, domestic rabbits and horses. The domesticated bovine farm animals that are raised for commodities are known as cattle. This chapter discusses in detail these types of livestock.

Chapter – Animal Husbandry

The branch of agriculture which is concerned with animals that are raised for milk, eggs, fiber and meat is referred to as animal husbandry. The important aspects of animal husbandry include animal breeding and animal feed. The chapter closely examines these key aspects of animal husbandry to provide an extensive understanding of the subject.

Chapter – Animal Farming Practices

Some of the common animal farming practices are cuniculture, dairying, sheep farming, intensive animal farming and pig farming. Cuniculture refers to the breeding and raising of rabbits for wool, fur or meat while dairying involves the activities related to producing, storing and distributing milk and its products. All these diverse practices of animal farming have been carefully analyzed in this chapter.

Chapter – Livestock Management

Some of the activities which fall under the umbrella of livestock management are livestock dehorning, livestock grazing, livestock branding, livestock health and disease management, and manure management. These diverse aspects of livestock management have been thoroughly discussed in this chapter.

**Misha Peralta**

# 1
# Livestock Production

The domesticated animals that are raised in an agricultural setting to produce labor and commodities such as meat, milk, eggs, fur, etc. are known as livestock. Some of the common types of livestock production system are grassland-based livestock production system and landless ruminant production system. This chapter has been carefully written to provide an easy understanding of these livestock production systems.

Livestock production contributes to 40 percent of the global value of agricultural output and supports the livelihoods and food security of almost one billion people, and is expanding rapidly. However, the problem of child labour in this sector is often ignored. It can also be stationary and larger scale. In many cases, it can be combined with farming. In fact, a typical farm operation may combine the tasks of crop production and harvesting, livestock rearing and handling, and manure disposal.

Some tasks often categorized as domestic chores contribute to livestock production such as collecting grass for cattle, cleaning out cowsheds and looking after small livestock for domestic consumption.

## Livestock Farming

Livestock farming is raising of animals for use or for pleasure. In this topic, the discussion of livestock includes both beef and dairy cattle, pigs, sheep, goats, horses, mules, asses, buffalo, and camels; the raising of birds commercially for meat or eggs (i.e., chickens, turkeys, ducks, geese, guinea fowl, and squabs) is treated separately.

An efficient and prosperous animal agriculture historically has been the mark of a strong, well-developed nation. Such an agriculture permits a nation to store large quantities of grains and other foodstuffs in concentrated form to be utilized to raise animals for human consumption during such emergencies as war or natural calamity. Furthermore, meat has long been known for its high nutritive value, producing stronger, healthier people.

Ruminant (cud-chewing) animals such as cattle, sheep, and goats convert large quantities of pasture forage, harvested roughage, or by-product feeds, as well as nonprotein nitrogen such as urea, into meat, milk, and wool. Ruminants are therefore extremely important; more than 60 percent of the world's farmland is in meadows and pasture. Poultry also convert feed efficiently into protein; chickens, especially, are unexcelled in meat and egg production. Milk is one of the most complete and oldest known animal foods. Cows were milked as early as

9000 BCE. Hippocrates, the Greek physician, recommended milk as a medicine in the 5th century BCE. Sanskrit writings from ancient India refer to milk as one of the most essential human foods.

# Classification of Livestock Production Systems

The systems classification aims at:

- Delineating and defining elements of a classification of livestock production systems.

- Quantitatively and qualitatively describing each livestock production system in terms of feed and livestock resources livestock commodities produced; production technology; product use and livestock functions; area covered; geographic locations; and human populations supported.

- Providing insights into the importance of livestock systems across world regions and agro-ecological zones and related trends in order to provide orientation to decision-makers involved in livestock development.

The results form a valuable basis for priority setting in AGA's new programme of work, which will be based on a systems approach starting from the 1996-97 biennium. They were originally used as the basis for the environmental impact assessment of an FAO-coordinated global multidonor study on interactions between livestock production systems and the environment. The results are useful in a general discussion of livestock development.

The classification covers the following animal species: cattle, buffalo, sheep, goat, pig and chicken. In geographic terms, systems are grouped according to the following regions: sub-Saharan Africa (SSA); Asia; Central and South America (CSA); West Asia and North Africa (WANA); Organisation for Economic Co-operation and Development (OECD) member countries, excluding Turkey, which was included in WANA; eastern Europe and Commonwealth of Independent States (CIS); and other developed countries (Israel and South Africa). The study covers 150 countries, comprising over 98 percent of the world production of the commodities concerned.

Livestock production systems are considered to be a subset of farming systems. These classifications are closer to typologies. No attempts at developing a classification of world livestock systems by using quantitative statistical methodologies (cluster analysis and related methodologies) could be located in the literature. This probably relates to the lack of appropriate data sets for such approaches on a global scale.

The classification criteria were limited to three: integration with crops, relation to land and agro-ecological zone. In addition, the landless system group was split into two - landless ruminant and landless monogastric - bringing the total number of systems to 11.

Using data from the FAO Information System for Agricultural Statistics (AGROSTAT), commodities or items were sequentially allocated to specific cells with defined attributes. At the first stage, national totals were assigned to one or more of the agro-ecological zones (AEZs) of the country in question using decision rules. For land-based production systems, for example, that is the proportion of arable land in each AEZ, and for landless production systems, the prorating factor became the population in each AEZ, in relative terms. The world's land surface was classified into ten AEZs.

The next stage was the overlay with classification criteria defining the attributes of the farming system, such as mixed versus grazing or mixed rain-fed versus mixed irrigated. This classification was based on decision rules related to the share of arable land as compared to grazing land and to the share of irrigated versus non-irrigated arable land.

Data for each item were then aggregated across specified AEZs to arrive at climatically broader and less numerous systems, for example, humid + subhumid tropics and subtropics. Data on different dimensions of a livestock production system were extracted from the item-oriented spreadsheets to produce system-descriptive spreadsheets.

Given the intrinsic weakness of the procedure of allocation to systems, in the case of large countries with diverse ecologies, subnational statistics were consulted to allocate the data manually to a production system for the major countries: China and India in Asia; Nigeria, the Sudan and Ethiopia in sub-Saharan Africa; and Brazil, Mexico and the United States in the Americas.

## Solely Livestock Production Systems

Livestock systems in which more than 90 percent of dry matter fed to animals comes from range-lands, pastures, annual forages and purchased feeds and less than 10 percent of the total value of production comes from non-livestock farming activities.

- Landless livestock production systems (LL): Subset of the solely livestock production systems in which less than 10 percent of the dry matter fed to animals is farm-produced and in which annual average stocking rates are above ten livestock units (LU) per hectare of agricultural land.

- Grassland-based systems (LG): Subset of solely livestock production systems in which more than 10 percent of the dry matter fed to animals is farm-produced and in which annual average stocking rates are less than ten LU per hectare of agricultural land.

## Mixed-farming Systems

Livestock systems in which more than 10 percent of the dry matter fed to animals comes from crop by-products or stubble or more than 10 percent of the total value of production comes from non-livestock farming activities.

- Rain-fed mixed-farming systems (MR): A subset of the mixed systems in which more than 90 percent of the value of non-livestock farm production comes from rain-fed land use.

- Irrigated mixed-farming systems (MI): A subset of the mixed systems in which more than 10 percent of the value of non-livestock farm production comes from irrigated land use.

## Description of Systems

This information is supplemented by a brief description of the main features of each system as well as the development paths along which these systems are evolving.

## Landless Livestock Production Systems

The developed countries dominate the picture of landless intensive production with more than half of total meat production as shown in figure. Asia is already contributing some 20 percent and eastern Europe 15 percent, with the latter recently in sharp decline.

1. Landless monogastric production system (LLM): This system is defined by the use of monogastric species, mainly chickens and pigs, where feed is introduced from outside the farm, thus separating decisions concerning feed use from those of feed production, and particularly of manure utilization on fields to produce feed and/or cash crops. This system is therefore open in terms of nutrient flow.

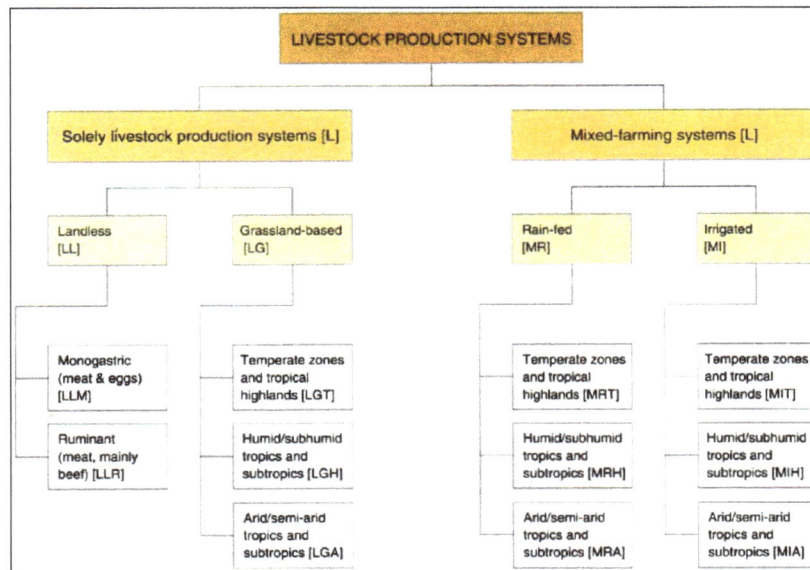

Classification of world livestock production systems.

Landless monogastric systems are found predominantly in OECD member countries with 52 percent of the total landless pork production and 58 percent of the landless poultry production globally. In the case of pig production, Asia is second, with 31 percent of the world total. For poultry, Central and South America follow, with 15 percent. To a large extent, this geographic distribution is determined by markets and consumption patterns in addition to levels of urbanization.

In Southeast and eastern Asia, this system is especially important. As much as 96 percent of the total pig-meat production in Asia occurs in China, Viet Nam and Indonesia. China, Thailand and Malaysia produce 84 percent of poultry meat. This is associated with fast economic growth and urbanization. The demand for monogastric meat is expected to grow from two- to fivefold between 1987 and 2006 from a base of 31 million tonnes, and a three- to tenfold increase is expected in the demand for eggs from 9 million tonnes. The prerequisites for development into large-scale vertically integrated production include the use of appropriate breeds and strains, feed quantity and quality, housing and disease control, as well as assured markets both at home and abroad. Landless poultry and pig production systems account for the majority of the output in developed countries and their share is rapidly increasing in developing countries given their high supply elasticity in the short term.

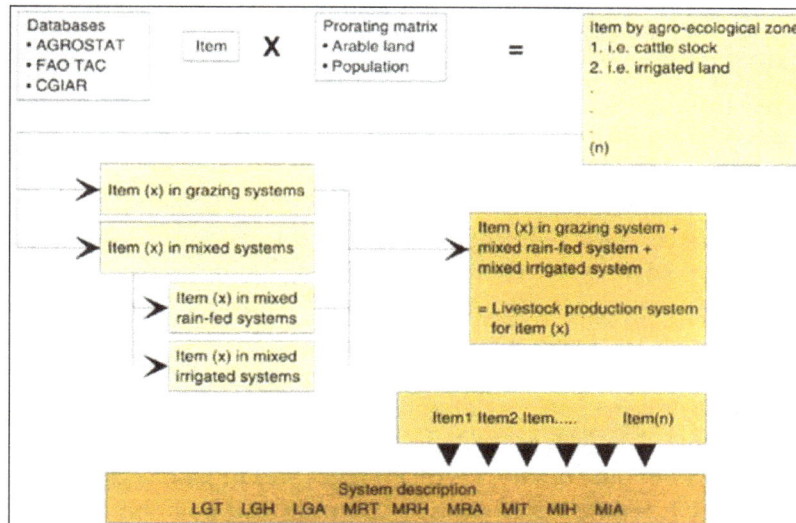

Flow of data processing for system quantification.

The system is typically competing with traditional land-based production systems for shares in the urban markets. It must be kept in mind that poultry and pork are close substitutes for beef and mutton, thus also interacting with the ruminant systems. In a broader sense, the demand for cereals created by these systems is also competing for land resources with land-based ruminant systems.

2. Landless ruminant production system (LLR): This production system is defined by the use of ruminant species, principally cattle, where feed is mainly introduced from outside the farm system. Landless ruminant production systems are highly concentrated in only a few regions of the world. In the case of cattle, they are almost exclusively found in eastern Europe and the CIS and in a few OECD member countries. Landless sheep production systems are only found in western Asia and northern Africa.

Typical cases are large-scale feedlots in the United States and in eastern Europe and the CIS, and veal production in parts of the KU. Intensive dairy operations in the same regions are more land-based because of the need to feed palatable fodder, which cannot be transported economically over long distances. In Asia, the intensive ruminant systems are typically found in buffalo and dairy cattle production units/colonies in India and Pakistan.

The LLR system is based almost exclusively on high-producing, specialized breeds and their crosses, which, nevertheless, have not been bred specifically for performance under "landless" conditions. With regard to milk production, the Holstein-Friesian breed is clearly the most important, and for beef production, English breeds predominate in the United States, while the large European dual-purpose breeds provide animals for fattening. The abundance of rangeland in the United States has led to the specialized production of calves from beef breeds for feedlot operations, while under European conditions these animals are a joint product together with milk, mainly from mixed systems.

The LLR system is highly capital-intensive, leading to substantial economies of scale. It is also feed-intensive and labour-extensive. Key efficiency parameters are daily weight gains and feed conversion, basically reflecting the efficient use of capital invested in infrastructure or in the form of lean animals and feeds. Weight gains are usually in the range of 1 to 1.5 kg per day, and feed conversion rates are about 8 to 10 kg of grains per kilogram of weight gain.

This system is closely linked to land-based systems that normally provide the young stock for landless systems. This constitutes an important difference from landless monogastric systems, in which replacement stock is produced within the same vertically integrated system.

Driven by population growth, the landless ruminant system is expected to continue to grow slowly in North America and southern Asia. On the other hand, it is expected to decline in the EU as production becomes more extensive in response to policies reducing agricultural support. In eastern Europe and the CIS, its importance is also declining and ruminant production in that part of the world is shifting to land-based and smaller-scale systems.

A growing market for grain-fed beef exists in Japan and the newly industrialized countries of Asia. The growth rate of this market will depend mainly on the evolution of the international price of cereals and the increased per caput incomes. This market will in part be supplied domestically and through imports from the United States, Canada, Australia and possibly South America.

## Grassland-based Livestock Production Systems

The importance of grassland-based systems in different world regions is shown in figure. Central and South America and the developed countries dominate the picture in terms of meat production, together accounting for more than three-quarters of the world's production.

1. Temperate zones and tropical highlands grassland-based system (LGT): In these areas, the grazing system is constrained by low temperatures. In the temperate zones, there are one or two months of mean temperatures, corrected to sea level, to below 5°C, whereas in the tropical highlands daily mean temperatures during the growing period are in the range of 5° to 20 °C.

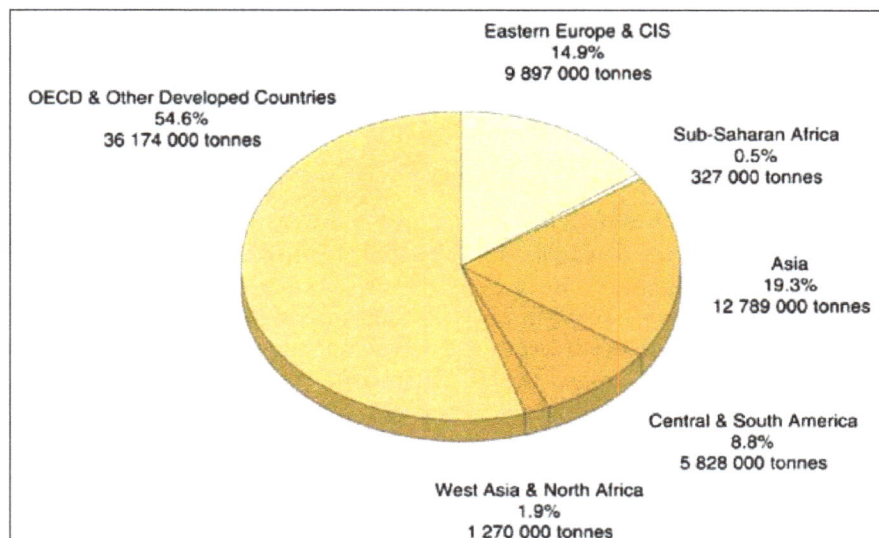

Total meat production of landless farming systems in different regions of the world.

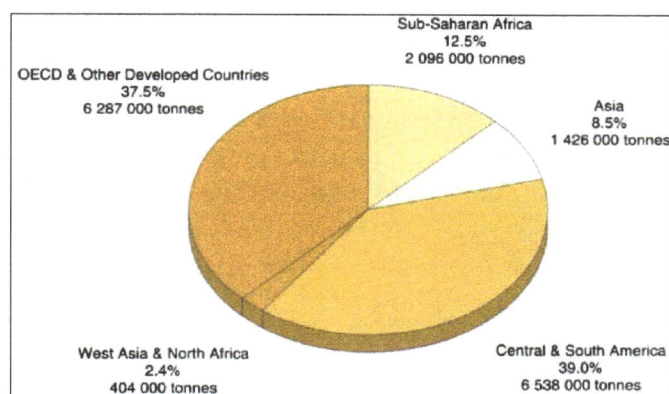

Total meat production of grazing systems in different regions of the world.

Locations in the tropical highlands comprise parts of the highlands of South America and eastern Africa, and in temperate zones they include southern Australia, New Zealand and parts of the United States, China and Mongolia.

Typical cases are Mongolia's steppe system, New Zealand's dairy and sheep enterprises, dairy systems close to Bogota, Colombia, and the South American camelid and sheep grazing systems in the altiplano of Peru and Bolivia. Extensive grazing systems are also found in parts of northwestern Pakistan involving sheep for mutton and wool and transhumant sheep on degraded high-altitude pasture in Nepal. Further cases are reported for Chinese merino wool sheep on communal grazing in Jilin Province and sheep ranching on grass-clover pastures of intensive animal production enterprises in the state of Oregon in the United States.

The regions in which the LGT system predominates have a combined human population of 190 million, which represents only 3.5 percent of the world total, and almost half of this population lives in Asia. In OECD member countries, far fewer people (14 million or 1.7 percent) use the LGT system, but they control more land and cattle per inhabitant than in the other regions.

Under the LGT system, product use varies widely, ranging from export-oriented New Zealand farmers, to South American farmers mainly producing for the domestic market, to Asian and African smallholders concerned with local markets and their own subsistence.

Market forces and environmental concerns are curbing the potential for intensification of this system. As a result, their global market share is declining vis-à-vis that of other production systems.

Since the LGT system is found mostly in marginal areas, its production potential in global terms is relatively low. In developing countries it tends to form a subsistence basis for certain groups of the population and its future role is seen more in providing employment for these groups than in making a major contribution to output and economic development. In developed countries, frequently with production surpluses, the production from these systems is declining in relation to other values and uses assigned to these land resources, such as recreational value, value as a wildlife and biological diversity reserve and the contribution to water conservation.

2. Humid and subhumid tropics and subtropics grassland-based system (LGH): The LGH system is defined as a grazing system found in regions with more than 180 days of growing period. It tends to be concentrated more in the subhumid zone, particularly in regions where access to markets

or, for agronomic reasons, crop production is limited. By definition, only very limited cropping is considered for subsistence.

The LGH system is found mostly in the tropical and subtropical lowlands of South America: in the llanos of Colombia and Venezuela as well as in the cerrados of Brazil. Dual-purpose milk-beef systems in the Mexican lowlands and estancias in Argentina are also typical cases of this system. In the African setting, many of the potentially suitable land resources are not used as a result of try-panosomiasis constraining livestock production. Outside Latin America, this system is important only in Australia because of its ample land resources in relation to its population.

Worldwide, the LGH system comprises about 190 million head of cattle, an important proportion of which are of the zebu breeds. In the subhumid and humid regions, cattle are clearly the dominant species, and in very high rainfall areas, such as the Amazon River delta and some parts of Queensland, Australia, buffaloes are also ranched. African hair sheep and dwarf goats are usually kept for local consumption only. In the subtropics, wool sheep are an important component of the system, for example, in Argentina, Uruguay, South Africa and Australia.

The LGH system produces approximately 6 million tonnes of beef and veal and 11 million tonnes of cow milk worldwide. By far the most important geographic region is Central and South America. The system is predominantly market-oriented.

Globally, 6 percent of the world's population (330 million people) lives in areas where the LGH production system predominates. Its importance in terms of sustaining the livelihood of rural populations is expected to decline as interaction with crop cultivation turns it into a mixed system. In rain-forest regions, efforts are being made to incorporate perennial tree crops, frequently into silvo-pastoral systems. In the savannahs, this system is being converted into a mixed-farming system by including annual crops, such as maize, soybeans and sorghum.

Improvements in road infrastructure and new technologies to allow the establishment of pastures with commercially worthwhile nurse crops are making the ley farming systems, involving rotations of crops and pasture, a potentially attractive pathway into mixed-farming systems.

3. Arid and semi-arid tropics and subtropics grassland-based system (LGA): The LGA system is defined as a land-based system in tropical and subtropical regions with a growing period of less than 180 days, and where grazing ruminants is the dominant form of land use.

This system is found under two contrasting socioeconomic frameworks: in sub-Saharan Africa and the Near East and North Africa regions, where it constitutes a traditional way of subsistence for an important part of the population, and in Australia, parts of western United States and southern Africa, where private enterprises utilize publicly or privately owned range resources for ranching purposes.

While in western Asia, northern Africa and sub-Saharan Africa, the LGA system is important for the livelihood of large sectors of the rural population, in developed countries it is extremely la-bour-extensive.

In sub-Saharan Africa, agropastoralism is the most important interface between livestock production and other agricultural production. In other regions these systems are interrelated with other livestock production systems that have access to better-quality feed and are closer to markets. In low-income countries without an export market, incentives to produce quality beef are weak. This, in turn, limits the attractiveness of livestock production stratification.

Globally, new roles are emerging for these rangelands, besides that of producing ruminant animal products. In developing countries, the pressure to expand crop production is increasing population pressure on the remaining rangelands. In developed countries, the utilization of these rangelands for animal production has often been subsidized through very low prices for grazing permits and public investments in irrigation.

## Rain-fed Mixed-farming Systems

Sub-Saharan Africa, West Asia and North Africa, and Central and South America are relatively unimportant in terms of meat production, whereas developed countries and Asia together contribute about 70 percent of the total meat production from mixed-farming systems.

1. Temperate zones and tropical highlands rain-fed system (MRT): This system is defined as a combination of rain-fed crop and livestock farming in temperate or tropical highland areas, in which crops contribute at least 10 percent of the value of total farm output.

The MRT system is found in two contrasting agro-ecozones of the world: it is the dominant system in most of North America, Europe and northeastern Asia, basically covering large strips of land north of the 30° northern latitude parallel, and it is found in the tropical highlands of eastern Africa (Ethiopia, Kenya, Uganda, Burundi, Rwanda) and in the Andean region of Latin America (Ecuador, Mexico).

The main common feature of these two regions is that low temperatures during all or part of the year limit and determine vegetation that is quite distinct from that found in tropical environments (e.g. C3 versus C4 grasses).

In the course of the development process, production technology in temperate locations evolved to higher specialization, more use of external inputs and more open systems. This has resulted in increasing negative externalities of these systems for the environment.

In most tropical MRT systems, production is less intensive, with livestock performing a series of functions in mixed systems: a continuous flow of cash income; means to concentrate nutrients for crops through manure; fuel; animal traction; a cash reserve for emergencies; and as a buffer to risks in crop production.

Globally, the MRT system is the most important source of animal products, providing 39 percent of the beef and veal production, 24 percent of the mutton production and 63 percent of the cow milk produced.

In developed countries, growing environmental concern, reduced protection of domestic production and increased international trade have led to a stagnation or even a reduction in output levels. In eastern Europe and the CIS, mixed systems will replace inefficient large-scale landless systems as these economies open up to the markets.

In Asia, this system has the potential for increased production and better use of feeds for intensified ruminant production, associated with reallocation of land-use patterns at the farm level.

In the highlands of sub-Saharan Africa and Latin America, production increases must come from a further intensification of crop-livestock land-use systems. The highlands should favour small-scale mechanization because of the high population density, continuous cultivation and relatively heavy soils.

2. Humid and subhumid tropics and subtropics rain-fed system (MRH): In the humid and subhumid regions of the tropics and subtropics, livestock production is based on mixed-farming systems. Given the range of socioeconomic conditions and soils and climates involved, this livestock system is very heterogeneous in its composition. It is found in all tropical regions of the world, mainly in developing countries. Parts of the southern United States are the only significantly developed regions included in this system. Typical cases are smallholder rice-buffalo systems in Southeast Asia or soybean-maize-pasture large-scale commercial operations in the Brazilian cerrados.

This system includes regions with especially difficult climatic conditions for livestock (high temperatures and high humidity). Adaptation of highly productive temperate breeds to these challenges has been notably poor. In many parts of Africa, trypanosomiasis constitutes an additional constraint to these systems. Particularly in African and Asian smallholder systems the local breeds are still widely used. In Latin America, Bos taurus cattle, sheep and goats were introduced some four centuries ago. Bos indicus cattle were introduced a few decades ago and have now replaced the earlier introduced cattle breeds in tropical areas.

In the African and Asian MRH system, the multiple roles of livestock have prevailed, particularly animal traction and manure. In Central and South America, this system caters to large domestic markets and, particularly in the case of Brazil, it is also linked to export markets. Under smallholder conditions, milk tends to be a more important output than meat.

The MRH system applies to approximately 14 percent of the global population. This ratio is particularly high in sub-Saharan Africa, where 41 percent of the region's population is associated with the system, and in Central and South America, where it is 35 percent. The system is replacing grazing systems in Africa and Latin America. In Africa, the process is mainly driven by population growth, and, in Central and South America, by economic development and technological innovations.

The main challenge in sub-Saharan Africa is finding ways to increase productivity. It is generally acknowledged that the biological potentials of mixed systems will be the key to productivity increases, and the expectation is that purchased feed inputs will be replaced by nutrients cycled within the system. In Asia, increased crop production - and hence feed availability is an important way of intensifying and maximizing animal productivity, however, the prospects for increasing cultivation area are limited.

In Latin America, low population density, a high degree of urbanization and relatively high per caput incomes have resulted in farming systems that are generally more oriented towards livestock production. In the tropical rain-forest regions, high resource-consuming systems were established, in some cases driven by policies and in other cases by poverty. Many of the policies that promoted wasteful utilization of these resources have been stopped in the process of structural adjustment.

3. Arid and semi-arid tropics and subtropics rain-fed system (MRA): The MRA system is a mixed-farming system in tropical and subtropical regions with a vegetation growth period of less than 180 days. The main restriction of this system is the low primary productivity of the land

resulting from low rainfall. The more severe the constraint, the less important crops become in the system and the more livestock take over as a primary income and subsistence source.

This system is important in the West Asia and North Africa region in parts of the Sahel (Burkina Faso, Nigeria), in large parts of India and in northeastern Thailand and eastern Indonesia, and less important in Central and South America. Given the low intensity of the system and the multiple purposes of livestock, the introduction of improved breeds has been limited. Globally, 11 percent of cattle and 14 percent of sheep and goats are found in this system. Small ruminants are particularly important in West Asia and North Africa under the MRA system.

As is the case in other largely smallholder systems, livestock have a range of simultaneous roles in this system, including animal traction, production of manure and use as a cash reserve, in addition to the production of meat and milk. Fuelwood is often scarce as a result of deforestation and range degradation, leading to the ever-increasing role of animals as providers of manure for fuel, in addition to means of transport.

While this system supports larger populations than any other grazing system, only 10 percent of the world's population is related to this system. Fifty-one percent of the population involved is in Asia, mainly India, and 24 percent is in the West Asia and North Africa region. There is close interaction with the LGA system. With increasing population pressure, the LGA system tends to evolve into mixed systems, mainly MRA, because of the greater caloric efficiency of cropping vis-à-vis ruminant production when land becomes scarce.

The outlook for this system is relatively similar to that for the LGA system. The resource base puts a clear ceiling on agricultural intensification. Low and variable response to inputs makes their use financially risky. Population growth in this setting is contributing to the overexploitation of the natural resource base, as traditional property rights cannot cope with the growing demands. Alternative development strategies and the reduction of population pressure on the resource base are key elements for the sustainable development of these regions.

## Irrigated Mixed-farming Systems

1. Temperate zones and tropical highlands mixed system (MIT): This system belongs to the group of land-based mixed systems of temperate and tropical highland regions. It is found particularly in the Mediterranean region (Portugal, Italy, Greece, Albania, Bulgaria) and in the Far East (the Democratic People's Republic of Korea, the Republic of Korea, Japan and parts of China), where plant growth is limited both by low temperatures in the cold season and reduced moisture availability during the vegetation period. The system's importance in tropical highlands is negligible.

Meat, milk and wool, the main outputs of this system, are mainly produced for the market. Manure is an issue only where animals are stabled, at least for certain periods of the day or the year.

About 10 percent of the global population lives in regions where this system is dominant. A large share belongs to developed countries with relatively high income levels and where agricultural trade is important. This system is clearly associated with very intensive agriculture in temperate regions with a high population density. With the outcome of the General Agreement on Tariffs and Trade (GATT) negotiations, it can be expected that these systems will be less and less viable, having to compete with more efficient rain-fed systems producing the same commodities. The expansion of international trade, and particularly the incorporation of southern European countries into the KU, has led to an increase in intensive production systems for off-season vegetables and fruits on the best irrigated land. The integration with livestock has been reduced, with ruminant grazing systems declining in absolute terms and being concentrated on the marginal sites.

2. Humid and subhumid tropics and subtropics mixed system (MIH); This is a mixed system in tropical and subtropical regions with growing seasons of more than 180 days, and in which the irrigation of crops is significant.

The MIH system is particularly important in Asia. High population densities require intensive crop production, and the irrigation of rice makes it possible to obtain more than two crops per year, even under conditions of very seasonal rainfall, substantially reducing yield variability as compared with the yield of upland rice or other rain-fed crops. In the past, animal production has been closely linked to the animal traction issue.

MIH systems throughout the world produce 13 million tonnes of pork (18 percent of global production), more than any other land-based tropical system. Among the tropical and subtropical systems, the MIH system is the one related to the largest population group, 990 million people, 97 percent of which are in Asia. Increasing labour productivity and relative affluence of farmers in this system are reflected in the more frequent use of tractors for cultivation. Manure is recycled on the fields. Ducks do well in this environment, as do pigs and poultry, which thrive on abundant crop residues. In addition, since large ruminants, such as buffaloes and, to a lesser extent, cattle, have little association with draught and transport, their output will be increased milk and meat for the market. Animals and intensive crop production in this ecological zone are an illustration of a successful and sustainable agricultural production system.

Competition for urban markets for livestock products is the main form of interaction with the landless monogastric system, both domestically and globally, through international trade.

3. Arid and semi-arid tropics and subtropics mixed system (MIA): This is a mixed system of arid and semiarid regions, in which irrigation makes year-round intensive crop production feasible. It is found in the Near East, South Asia, North Africa, western United States and Mexico.

Typical cases are alfalfa/maize-based intensive dairy systems in California, Israel and Mexico; small-scale buffalo milk production in Pakistan; and animal traction-based cash-crop production in Egypt and Afghanistan. Cattle and buffaloes for milk and animal traction are the main ruminant resources, although sheep and goats are important where marginal rangelands are available in addition to irrigated land. In the MIA system, pigs are kept only in the Far East; they are virtually non-existent in West Asia and North Africa, largely for cultural reasons (Islamic and Jewish religions). The main introduced breeds are dairy cattle to supply milk to large urban centres. Under good management conditions, intensive dairy schemes have been quite successful in hot but dry environments. Some of the world's highest lactation yields are achieved in the MIA system in Israel and California. The traditional smallholder MIA system in Asia relies heavily on buffaloes for milk production.

In the traditional MIA system, irrigated crop production is the main source of income, with livestock playing a very secondary role. This is generally reflected in rather extensive management of livestock enterprises. The MIA system is predominant in regions that are home to over 750 million people, two-thirds of them in Asia and one-third in West Asia and North Africa.

The main interaction with other systems occurs through the international market, particularly for milk and dairy products.

The MIA system makes an important contribution to food availability and employment in semi-arid and arid regions. Irrigation allows increased fodder production as a by-product or part of crop rotation, which reduces the feed deficit. The improved feed base and utilization promotes intensification and commercialization of livestock production, especially in areas with good market access.

# Livestock based Integrated Farming System

Livestock based integrated farming system is one of the rising agriculture systems for the northeastern region. The practice of this type of farming system has been continued in this region in a traditional way from time immemorial. The basic principles of the farming system are productive recycling of farm wastes. Different subsystems work together in integrated farming system resulting in a greater total productivity than the sum of their individual production. Fish-Livestock along with Livestock-Crop farming is the major concept in Livestock based integrated farming system.

## Fish-livestock Farming Systems

Fish-livestock farming systems are recognized as highly assured technology where predetermined quantum of livestock waste obtained by rearing the live stock in the pond area is applied in pond to raise the fish crop without any other additional supply of nutrients. The main potential linkages between livestock and fish production concern use of nutrients, particularly reuse of livestock manures for fish production. The term nutrients mainly refer to elements such as nitrogen (N) and phosphorus (P) which function as fertilizers to stimulate natural food webs rather than conventional livestock nutrition usage such as feed ingredients. Both production and processing of livestock generate by-products that can be used for aquaculture. Direct use of livestock production wastes is the most widespread and conventionally recognized type of integrated farming. Production wastes include manure, urine and spilled feed; and they may be used as fresh inputs or be processed in some way before use.

Based on the type of livestock used for integration there are many combinations in livestock-fish systems.

## Cattle-fish Culture

Manuring of fish pond by using cow dung is one of the common practices all-over the world. A healthy cow excretes over 4,000-5,000 kg dung, 3,500-4,000 lt urine on an annual basis. Manuring with cow dung, which is rich in nutrients results in increase of natural food organism and bacteria in fishpond. A unit of 5-6 cows can provide adequate manure for 1 ha of pond. In addition to 9,000 kg of milk, about 3,000-4,000 kg fish/ha/year can also be harvested with such integration.

Cowshed should be built close to fishpond to simplify handling of cow manure. A cow requires about 7,000-8,000 kg of green grass annually. Grass carp utilizes the left over grasses, which are about 2,500 kg. Fish also utilize the fine feed which consists of grains wasted by cows. In place of raw cow dung, biogas slurry could be used with equally good production. Twenty to thirty thousand kg of biogas slurry are recycled in 1 ha water area to get over 4000 kg of fish without feed or any fertilizer application.

## Pig-fish System

The waste produced by 30-40 pigs is equivalent to 1 tonne of ammonium sulphate. Exotic breeds like White Yorkshire, Landrace and Hampshire are reared in pig-sty near the fish pond. Depending on the size of the fishponds and their manure requirements, such a system can either be built on the bund dividing two fishponds or on the dry-side of the bund. Pigsties, however, may also be constructed in a nearby place where the urine and dung of pigs are first allowed to the oxidation tanks (digestion chambers) of biogas plants for the production of methane for household use. The liquid manure (slurry) is then discharged into the fishponds through small ditches running through pond bunds. Alternately, the pig manure may be heaped in localized places of fishponds or may be applied in fishponds by dissolving in water.

Pig dung contains more than 70 percent digestible feed for fish. The undigested solids present in the pig dung also serve as direct food source to tilapia and common carp. A density of 40 pigs has been found to be enough to fertilize a fish pond of one hectare area. The optimum dose of pig manure per hectare has been estimated as five tonnes for a culture period of one year. Fish like grass carp, silver carp and common carp (1:2:1) are suitable for integration with pigs.

Pigs attain slaughter maturity size (60-70 kg) with in 6 months and give 6-12 piglets in every litter. Their age at first maturity ranges from 6-8 months. Fish attain marketable size in a year. Final harvesting is done after 12 months of rearing. It is seen that a fish production of 3,000 kg/ha could be achieved under a stocking density of 6,000 fish fingerlings/ha in a culture period of six months.

## Poultry-fish Culture

Poultry raising for meat (broilers) or eggs (layers) can be integrated with fish culture to reduce costs on fertilizers and feeds in fish culture and maximize benefits. Poultry can be raised over or adjacent to the ponds and the poultry excreta recycled to fertilize the fishponds. Poultry housing, when constructed above the water level using bamboo poles would fertilize fishponds directly.In fish poultry integration, birds housed under intensive system are considered best. Birds are kept in confinement with no access to outside. Deep litter is well suited for this type of farming. About 6-8 cm thick layer prepared from chopped straw, dry leaves, saw dust or groundnut shell is sufficient.

Poultry dung in the form of fully built up dip litter contains: 3% nitrogen, 2% phosphate and 2% potash, therefore it acts as a good fertilizer which helps in producing fish feed i.e. phytoplankton and zooplankton in fish pond. So application of extra fertilizer to fish pond for raising fish is not needed. This cuts the cost of fish production by 60%. In one year 25-30 birds can produce 1 tonne dip litter and based on that it is found that 500-600 birds are enough to fertilize 1 ha water spread area for good fish production. Daily at the rate of 50 kg/ha water spread area poultry dung is applied to the fish pond. When phytoplanktonic bloom is seen over the surface water of fish pond then application of poultry dung to the pond should immediately be suspended. Poultry-fish integration also maximizes the use of space; saves labour in transporting manure to the ponds and the poultry house is more hygienic and water needed for poultry husbandry practice can get from fish pond.

## Duck-fish Culture

A fish-pond being a semi-closed biological system with several aquatic animals and plants,provides excellent disease-free environment for ducks.In return ducks consume juvenile frogs, tadpoles and dragonfly, thus making a safe environment for fish. Duck dropping goes directly in pond, which in turn provides essential nutrients to stimulate growth of natural food.This has two advantages, there is no loss of energy and fertilization is homogeneous.

It is highly profitable as it greatly enhances the animal protein production in terms of fish and duck per unit area. Ducks are known as living manuring machines. The duck dropping contain 25 per cent organic and 20 percent inorganic substances with a number of elements such as carbon, phosphorus, potassium, nitrogen, calcium, etc. Hence, it forms a very good source of fertilizer in fish ponds for the production of fish food organisms. Besides manuring, ducks eradicate the unwanted insects, snails and their larvae which may be the vectors of fish pathogenic organisms and water-borne disease-causing organisms infecting human beings. Further, ducks also help in releasing nutrients from the soil of ponds, particularly when they agitate the shore areas of the pond.

For duck-fish culture, ducks may be periodically allowed to range freely, or may be put in screened resting places above the water. Floating pens or sheds made of bamboo splits may also be suspended in the pond to allow uniform manuring. The ducks may be stocked in these sheds at the rate of 15 to 20/m². It is better if the ducks are left in ponds only until they reach marketable size. Depending on the growth rate of ducks, they may be replaced once in two to three months. About 15-20 days old ducklings are generally selected. The number of ducks may be between 100 and 3,000/ha depending on the duration of fish culture and the manure requirements.

For culturing fish with ducks, it is advisable to release fish fingerlings of more than 10 cm size, otherwise the ducks may feed on the fingerlings. The stocking density of fingerlings also depends on the size of pond and number of ducks released in it. As the nitrogen-rich duck manure enhances both phyto and zooplankton production, phytoplankton-feeding silver carp and zooplankton-feeding catla and common carp are ideal for duck-fish culture. The fish rearing period is generally kept as one year and under a stocking density of 20,000/ha, a fish production of 3,000-4,000 kg/ha/year has been obtained in duck-fish culture. In addition to this, eggs and duck-meat are also obtained in good quantity on an annual basis.

## Livestock-crop Production System

An "integrated crop-livestock system" is a form of mixed production that utilizes crops and livestock in a way that they can complement one another through space and time. The backbone of an integrated system is the herd of ruminants (animals like sheep, goats or cattle), which graze a pasture to build up the soil. Eventually, sufficient soil organic matter builds up to the point where crops can be supported. Animal can also be used for farm operations and transport. While crop residues provide fodder for livestock and grain provides supplementary feed for productive animals.

Animals play key and multiple roles in the functioning of the farm, and not only because they provide livestock products (meat, milk, eggs, wool, and hides) or can be converted into prompt cash in times of need. Animals transform plant energy into useful work: animal power is used for ploughing, transport and in activities such as milling, logging, road construction,marketing, and water lifting for irrigation. Animals also provide manure and other types of animal waste. Animal excreta have two crucial roles in the overall sustainability of the system:

- Improving nutrient cycling: Excreta contain several nutrients (including nitrogen, phosphorus and potassium) and organic matter, which are important for maintaining soil structure and fertility. Through its use, production is increased while the risk of soil degradation is reduced.

- Providing energy: Excreta are the basis for the production of biogas and energy for household use (e.g. cooking, lighting) or for rural industries (e.g.powering mills and water pumps). Fuel in the form of biogas or dung cakes can replace charcoal and wood.

One key advantage of crop-livestock production systems is that livestock can be fed on crop residues and other products that would otherwise pose a major waste disposal problem. For example, livestock can be fed on straw, damaged fruits, grains and household wastes. Integration of livestock and crop allows nutrients to be recycled more effectively on the farm. Manure itself is a valuable fertilizer containing 8 kg of nitrogen, 4kg of phosphorus and 16 kg of potassium per tonne. Adding manure to the soil not only fertilizes it but also improves its structures and water retention capacity. It is also opined that where livestock are used to graze, the vegetation under plantations of coconut, oil palm and rubber, as in Malaysia, the cost of weed control can be dramatically reduced, sometimes by as much as 40 percent. In Colombia sheep are sometimes used to control weeds in sugarcane. Draught animal power is widely used for cultivation, transportation, water lifting and powering food processing equipment.

## Overall Advantages of Integrated Farming System

- Productivity: IFS provides an opportunity to increase economic yield per unit area per unit time by virtue of intensification of crop and allied enterprises especially for small and marginal farmers.

- Profitability: Cost of feed for livestock is about 65-75% of total cost of production; however use of waste material and their byproduct reduces the cost of production, conversely it is same for the crop production as fertilizer requirement for crop is made available from animal excreta no extra fertilizer is required to purchase from out side farm as a result the benefit cost ratio increases and purchasing power of farmers improves thereby.

- Sustainability: In IFS, subsystem of one waste material or byproduct works as an input for the other subsystem and their byproduct or inputs are organic in nature thus providing an opportunity to sustain the potentiality of production base for much longer periods as compare to monoculture farming system.

- Balanced Food: All the nutrient requirements of human are not exclusively found in single food,to meet such requirement different food staffs have to be consumed by farmers. Such requirement can be fulfilled by adopting IFS at farmer level, enabling different sources of nutrition.

- Environmental Safety: In IFS waste materials are effectively recycled by linking appropriate components, thus minimize environment pollution.

- Recycling: Effective recycling of product, byproducts and waste material in IFS is the corner stone behind the sustainability of farming system under resource poor condition in rural area.

- Income Rounds the year: Due to interaction of enterprises with crops, eggs, meat and milk, provides flow of money round the year amongst farming community.

- Saving Energy: Cattle are used as a medium of transportation in rural area more over cow dung is used as such a burning material for cooking purpose or utilized to generate biogas thereby reducing the dependency on petrol/diesel and fossil fuel respectively, taping the available source within the farming system, to conserve energy.

- Meeting Fodder crisis: Byproduct and waste material of crop are effectively utilized as a fodder for livestock (Ruminants) and product like grain, maize are used as feed for monogastric animal (pig and poultry).

- Employment Generation: Combining crop with livestock enterprises would increase the labour requirement significantly and would help in reducing the problems of under employment to a great extent IFS provide enough scope to employ family labour round the year.

# Extensive Livestock Production Systems

Extensive livestock production systems are systems in which animals are kept free-range for part or all of their production cycle.

A Herdsman inspects his cattle, South Africa.

Cattle kept in a 'kraal', South Africa.

## Commercial Ruminant Production

The principal infectious disease threats are vector-borne diseases, reproductive diseases and, in most parts of Africa depending on the area, wildlife/livestock interface diseases. The main challenges for animal health management are:

- Sufficient observation to ensure early diagnosis of problems.

- Logistics of rounding up for observation, vaccination and treatment.

- Applying the necessary level of biosecurity to prevent uncontrolled access to pastures and animals.

## Production on Communal Grazing

In addition to the above diseases and challenges, additional challenges to animal health management are:

- Failure of treatment, e.g. prophylactic treatment against parasites, because not all the animals are treated owing to different levels of commitment amongst owners, resulting in high levels of pasture contamination.

- Build-up of pathogens that may affect production and reproduction (Brucella spp., Mycobacterium bovis, Salmonella) as a result of limited availability of land.

- Complete lack of control over access to the land by people, newly introduced animals and sometimes wildlife, resulting in exposure to a wide range of new pathogens including fatal diseases like MCF.

- Uncontrolled movement of animals.

## Extensive Pig Production

Most extensive pig production in sub-Saharan Africa occurs in traditional systems in which pigs are never confined, confined only at night or seasonally to protect crops, and find most of their own food.

Scavenging pigs, Ghana.

Pigs kept extensively are free from the majority of infectious diseases that commonly occur in confined pigs, such as PRRS, porcine multisystemic wasting syndrome, porcine pleuropneumonia, enzootic pneumonia, swine dysentery, procine proliferative enteropathy, Glässer's disease to name a few), but are more prone to ASF and CSF because they may be in contact with wild pigs, carcasses of pigs that have died, and pigs from other herds. Large populations of free-ranging pigs can become a reservoir in which the viruses of those diseases can circulate for long periods of time, probably indefinitely. They also often have access to human waste and in that way ingest the eggs of the pig tapeworm, Taenia solium, and become infested with the cysts formed by the larvae that migrate mainly to striated muscle. Ingestion of raw or undercooked pork containing cysts by humans, the definitive host, enables the cycle to be completed. There are no known clinical effects on the pig, and the effects of human taeniasis are generally mild, but neurocysticercosis caused by pig tapeworm, resulting from accidental ingestion of the eggs by humans and migration of the larvae to the brain, is a serious zoonosis recently recognised by WHO as a 'major neglected disease' and a major cause of epilepsy in sub-Saharan Africa and many of the poorest countries elsewhere where free-ranging pigs are raised and hygiene standards are low. Pork that is infested with cysts is not able to be sold in the formal market.

The main challenges for animal health management in extensively kept pigs are:

- Lack of observation leading to delayed diagnosis of serious diseases.

- Lack of investment in pig health by their owners.

- Lack of awareness by the owners about diseases like porcine cysticercosis.

- Uncontrolled movement of pigs.

- Difficulty in applying preventive or control measures.

In compliance with consumer demands for pigs to be kept under more natural conditions, pigs in outdoor systems are becoming more frequent in developed countries, for example member states of the European Union, with pressure on other pig-producing countries to follow suit. These systems differ markedly from the traditional extensive systems in that the pigs are fed, the property is fenced and there is likely to be some health management by the owners. Many of the diseases to which confined pigs are prone will be eliminated because pathogens are less concentrated and may be destroyed by exposure to the elements. However, increases in helminths including large roundworm (Ascaris suum) have been observed. Although the origin of the outbreak of CSF in UK in 2000 was never confirmed, it was first diagnosed in outdoor sows and it appeared that the disease may have been present for some time before it was noticed. It was also surmised that the first infection could have been due to infected material being fed by a passer-by to outdoor pigs, although this was of course impossible to prove.

## Extensive Poultry and Ostrich Production

There is inevitably contact between poultry kept out of doors and wild birds, which can enhance the chances of infection with Newcastle disease and avian influenza. As in all extensive systems, it is more likely that disease, unless it causes high mortality, will not be noticed and will be able to spread throughout the flock before it is noticed. This is particularly true in the case of subclinical infection, which apparently occurred in the 2011 outbreak of H5N2 avian influenza in South Africa. Like outdoor pigs, outdoor poultry flocks are spared the diseases of intensification that can occur in housed birds.

## Nomadic/Pastoralist Systems

Nomadic systems involve cattle, sheep, goats and camels, as the other commonly kept species of livestock are not susceptible to herding and therefore to moving about with their owners. The movement of herds varies from seasonal movement over relatively short distances to take advantage of grazing that is only seasonally available in another area to long-distance movements that may involve crossing the borders of one or more countries. Nomadism is a major reason why diseases like foot and mouth disease, peste des petits ruminants (PPR) and sheep- and goatpox and camelpox have proven impossible to eradicate, because flocks regularly come into contact with new flocks and infections can be exchanged. The main challenges for animal health management are:

- Mingling of herds and flocks that originated from different areas at water points, in grazing areas and at markets.

- Inconsistent or no access to animal health services.

Milking free-range cattle, Kenya.

On the other hand, nomadic livestock owners are usually strongly dependent on their animals for their livelihoods and will often take the best care of them that they can based on a fund of traditional knowledge.

## References

- Agriculture, areas, lang--en/index, WCMS: ilo.org,Retrieved 20 August, 2019

- Livestock-farming, topic: britannica.com, Retrieved 21 June, 2019

- Income-generation-and-nutritional-security-through-livestock-based-integrated-farming-system, livestock: vikaspedia.in, Retrieved 22 July, 2019

- Production-extensive, AHM, sites: afrivip.org. Retrieved 23 August, 2019

# 2
# Types of Livestock

Some of the diverse types of livestock are cattle, domestic sheep, domestic pigs, goats, donkeys, camels, domestic rabbits and horses. The domesticated bovine farm animals that are raised for commodities are known as cattle. This chapter discusses in detail these types of livestock.

## Cattle

Cattle is domesticated bovine farm animals that are raised for their meat, milk, or hides or for draft purposes. The animals most often included under the term are the Western or European domesticated cattle as well as the Indian and African domesticated cattle. However, certain other bovids such as the Asian water buffalo, the Tibetan yak, the gayal and banteng of Southeast Asia, and the plains bison of North America have also been domesticated or semidomesticated and are sometimes considered to be cattle.

Ankole-Watusi cattle.

In the terminology used to describe the sex and age of cattle, the male is first a bull calf and if left intact becomes a bull; if castrated he becomes a steer and in about two or three years grows to an ox. The female is first a heifer calf, growing into a heifer and becoming a cow. Depending on the breed, mature bulls weigh 450–1,800 kg (1,000–4,000 pounds) and cows 360–1,100 kg (800–2,400 pounds). Males retained for beef production are usually castrated to make them more docile on the range or in feedlots; with males intended for use as working oxen or bullocks, castration is practiced to make them more tractable at work. The use of cattle as commodities has been a point of philosophical contention throughout history, particularly regarding the raising of animals for food. Such issues are compounded by modern concerns about the ethics of industrial factory farming and the contribution of commercial meat production to global warming.

Hereford cow and calf.

All modern domestic cattle are believed to belong to the species Bos taurus (European breeds such as Shorthorn and Jersey) or Bos indicus (Zebu breeds such as Brahman) or to be crosses of these two (such as Santa Gertrudis). Breeds as they are known today did not always exist, and many are of recent origin. The definition of a breed is difficult and inexplicit, although the term is commonly used and, in practice, well understood. It may be used generally to connote animals that have been selectively bred for a long time so as to possess distinctive identity in colour, size, conformation, and function, and these or other distinguishing characteristics are perpetuated in their progeny. Breeds have been established by generations of breeders aiming at the attainment and preservation of a particular type with its identifying characteristics. This is accomplished by working on the principle of "like begets like." It is only in relatively recent times that the science of genetics, and particularly population genetics, has contributed to breeding.

Jersey cow.

There are many old established breeds in continental Europe—for example, the Charolais and Normande of France, the Holstein-Friesian of the Netherlands, and many others—but British breeds are of particular interest because of their influence in building up the vast herds that supply so much beef and dairy in other countries around the world.

# Domestic Sheep

Domestic sheep (Ovis aries) are quadrupedal, ruminant mammals typically kept as livestock. Like most ruminants, sheep are members of the order Artiodactyla, the even-toed ungulates. Although the name sheep applies to many species in the genus Ovis, in everyday usage it almost always refers to Ovis aries. Numbering a little over one billion, domestic sheep are also the most numerous species of sheep. An adult female sheep is referred to as a ewe, an intact male as a ram or occasionally a tup, a castrated male as a wether, and a younger sheep as a lamb.

Sheep are most likely descended from the wild mouflon of Europe and Asia. One of the earliest animals to be domesticated for agricultural purposes, sheep are raised for fleeces, meat (lamb, hogget or mutton) and milk. A sheep's wool is the most widely used animal fiber, and is usually harvested by shearing. Ovine meat is called lamb when from younger animals and mutton when from older ones in Commonwealth countries, and lamb in the United States (including from adults). Sheep continue to be important for wool and meat today, and are also occasionally raised for pelts, as dairy animals, or as model organisms for science.

Sheep husbandry is practised throughout the majority of the inhabited world, and has been fundamental to many civilizations. In the modern era, Australia, New Zealand, the southern and central South American nations, and the British Isles are most closely associated with sheep production.

Sheepraising has a large lexicon of unique terms which vary considerably by region and dialect. Use of the word sheep began in Middle English as a derivation of the Old English word scēap; it is both the singular and plural name for the animal. A group of sheep is called a flock, herd or mob. Many other specific terms for the various life stages of sheep exist, generally related to lambing, shearing, and age.

Being a key animal in the history of farming, sheep have a deeply entrenched place in human culture, and find representation in much modern language and symbology. As livestock, sheep are most often associated with pastoral, Arcadian imagery. Sheep figure in many mythologies—such as the Golden Fleece—and major religions, especially the Abrahamic traditions. In both ancient and modern religious ritual, sheep are used as sacrificial animals.

The exact line of descent between domestic sheep and their wild ancestors is unclear. The most common hypothesis states that *Ovis aries* is descended from the Asiatic (*O. orientalis*) species of mouflon. Sheep were among the first animals to be domesticated by humankind (although the domestication of dogs may have taken place more than 20,000 years earlier); the domestication date is estimated to fall between 11,000 and 9,000 B.C in Mesopotamia. The rearing of sheep for secondary products, and the resulting breed development, began in either southwest Asia or western Europe. Initially, sheep were kept solely for meat, milk and skins. Archaeological evidence from statuary found at sites in Iran suggests that selection for woolly sheep may have begun around 6000 BC, and the earliest woven wool garments have been dated to two to three thousand years later.

Sheep husbandry spread quickly in Europe. Excavations show that in about 6000 BC, during the Neolithic period of prehistory, the Castelnovien people, living around Châteauneuf-les-Martigues near present-day Marseille in the south of France, were among the first in Europe to keep domestic

## Breeds

Sheep being judged for adherence to their breed standard, and being
held by the most common method of restraint.

The domestic sheep is a multi-purpose animal, and the more than 200 breeds now in existence
were created to serve these diverse purposes. Some sources give a count of a thousand or more
breeds, but these numbers cannot be verified, according to some sources. However, several hundred breeds of sheep have been identified by the Food and Agriculture Organization of the UN
(FAO), with the estimated number varying somewhat from time to time: e.g. 863 breeds as of 1993,
1314 breeds as of 1995 and 1229 breeds as of 2006. (These numbers exclude extinct breeds, which
are also tallied by the FAO.) For the purpose of such tallies, the FAO definition of a breed is "either
a subspecific group of domestic livestock with definable and identifiable external characteristics
that enable it to be separated by visual appraisal from other similarly defined groups within the
same species or a group for which geographical and/or cultural separation from phenotypically
similar groups has led to acceptance of its separate identity." Almost all sheep are classified as
being best suited to furnishing a certain product: wool, meat, milk, hides, or a combination in a
dual-purpose breed. Other features used when classifying sheep include face color (generally white
or black), tail length, presence or lack of horns, and the topography for which the breed has been
developed. This last point is especially stressed in the UK, where breeds are described as either
upland (hill or mountain) or lowland breeds. A sheep may also be of a fat-tailed type, which is a dual-purpose sheep common in Africa and Asia with larger deposits of fat within and around its tail.

The Barbados Blackbelly is a hair sheep breed of Caribbean origin.

Breeds are often categorized by the type of their wool. Fine wool breeds are those that have wool of great crimp and density, which are preferred for textiles. Most of these were derived from Merino sheep, and the breed continues to dominate the world sheep industry. Downs breeds have wool between the extremes, and are typically fast-growing meat and ram breeds with dark faces. Some major medium wool breeds, such as the Corriedale, are dual-purpose crosses of long and fine-wooled breeds and were created for high-production commercial flocks. Long wool breeds are the largest of sheep, with long wool and a slow rate of growth. Long wool sheep are most valued for crossbreeding to improve the attributes of other sheep types. For example: the American Columbia breed was developed by crossing Lincoln rams (a long wool breed) with fine-wooled Rambouillet ewes.

Coarse or carpet wool sheep are those with a medium to long length wool of characteristic coarseness. Breeds traditionally used for carpet wool show great variability, but the chief requirement is a wool that will not break down under heavy use (as would that of the finer breeds). As the demand for carpet-quality wool declines, some breeders of this type of sheep are attempting to use a few of these traditional breeds for alternative purposes. Others have always been primarily meat-class sheep.

The Lička pramenka is a sheep breed of Croatian origin.

A minor class of sheep are the dairy breeds. Dual-purpose breeds that may primarily be meat or wool sheep are often used secondarily as milking animals, but there are a few breeds that are predominantly used for milking. These sheep produce a higher quantity of milk and have slightly longer lactation curves. In the quality of their milk, the fat and protein content percentages of dairy sheep vary from non-dairy breeds, but lactose content does not.

A last group of sheep breeds is that of fur or hair sheep, which do not grow wool at all. Hair sheep are similar to the early domesticated sheep kept before woolly breeds were developed, and are raised for meat and pelts. Some modern breeds of hair sheep, such as the Dorper, result from crosses between wool and hair breeds. For meat and hide producers, hair sheep are cheaper to keep, as they do not need shearing. Hair sheep are also more resistant to parasites and hot weather.

With the modern rise of corporate agribusiness and the decline of localized family farms, many breeds of sheep are in danger of extinction. The Rare Breeds Survival Trust of the UK lists 22 native breeds as having only 3,000 registered animals (each), and The Livestock Conservancy lists 14 as either "critical" or "threatened". Preferences for breeds with uniform characteristics and fast growth have pushed heritage (or heirloom) breeds to the margins of the sheep industry. Those that remain are maintained through the efforts of conservation organizations, breed registries, and individual farmers dedicated to their preservation.

## Diet

Sheep are herbivorous mammals. Most breeds prefer to graze on grass and other short roughage, avoiding the taller woody parts of plants that goats readily consume. Both sheep and goats use their lips and tongues to select parts of the plant that are easier to digest or higher in nutrition. Sheep, however, graze well in monoculture pastures where most goats fare poorly.

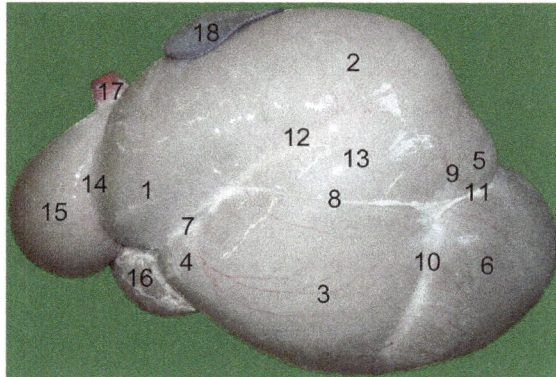

Ruminant system of a sheep.

Like all ruminants, sheep have a complex digestive system composed of four chambers, allowing them to break down cellulose from stems, leaves, and seed hulls into simpler carbohydrates. When sheep graze, vegetation is chewed into a mass called a bolus, which is then passed into the rumen, via the reticulum. The rumen is a 19- to 38-liter (5 to 10 gallon) organ in which feed is fermented. The fermenting organisms include bacteria, fungi, and protozoa. (Other important rumen organisms include some archaea, which produce methane from carbon dioxide.) The bolus is periodically regurgitated back to the mouth as cud for additional chewing and salivation. After fermentation in the rumen, feed passes into the reticulum and the omasum; special feeds such as grains may bypass the rumen altogether. After the first three chambers, food moves into the abomasum for final digestion before processing by the intestines. The abomasum is the only one of the four chambers analogous to the human stomach, and is sometimes called the "true stomach".

Other than forage, the other staple feed for sheep is hay, often during the winter months. The ability to thrive solely on pasture (even without hay) varies with breed, but all sheep can survive on this diet. Also included in some sheep's diets are minerals, either in a trace mix or in licks. Feed provided to sheep must be specially formulated, as most cattle, poultry, pig, and even some goat feeds contain levels of copper that are lethal to sheep. The same danger applies to mineral supplements such as salt licks.

## Grazing Behavior

Sheep follow a diurnal pattern of activity, feeding from dawn to dusk, stopping sporadically to rest and chew their cud. Ideal pasture for sheep is not lawnlike grass, but an array of grasses, legumes and forbs. Types of land where sheep are raised vary widely, from pastures that are seeded and improved intentionally to rough, native lands. Common plants toxic to sheep are present in most of the world, and include (but are not limited to) cherry, some oaks and acorns, tomato, yew, rhubarb, potato, and rhododendron.

Sheep grazing on public land (Utah, 2009).

## Effects on Pasture

Sheep are largely grazing herbivores, unlike browsing animals such as goats and deer that prefer taller foliage. With a much narrower face, sheep crop plants very close to the ground and can overgraze a pasture much faster than cattle. For this reason, many shepherds use managed intensive rotational grazing, where a flock is rotated through multiple pastures, giving plants time to recover. Paradoxically, sheep can both cause and solve the spread of invasive plant species. By disturbing the natural state of pasture, sheep and other livestock can pave the way for invasive plants. However, sheep also prefer to eat invasives such as cheatgrass, leafy spurge, kudzu and spotted knapweed over native species such as sagebrush, making grazing sheep effective for conservation grazing. Research conducted in Imperial County, California compared lamb grazing with herbicides for weed control in seedling alfalfa fields. Three trials demonstrated that grazing lambs were just as effective as herbicides in controlling winter weeds. Entomologists also compared grazing lambs to insecticides for insect control in winter alfalfa. In this trial, lambs provided insect control as effectively as insecticides.

## Behavior

Sheep showing flocking behavior during a sheepdog trial.

Flock of sheep in South Africa.

## Flock Behavior

Sheep are flock animals and strongly gregarious; much sheep behavior can be understood on the basis of these tendencies. The dominance hierarchy of sheep and their natural inclination to follow a leader to new pastures were the pivotal factors in sheep being one of the first domesticated livestock species. Furthermore, in contrast to the red deer and gazelle (two other ungulates of primary importance to meat production in prehistoric times), sheep do not defend territories although they do form home ranges. All sheep have a tendency to congregate close to other members of a flock, although this behavior varies with breed, and sheep can become stressed when separated from their flock members. During flocking, sheep have a strong tendency to follow, and a leader may simply be the first individual to move. Relationships in flocks tend to be closest among related sheep: in mixed-breed flocks, subgroups of the same breed tend to form, and a ewe and her direct descendants often move as a unit within large flocks. Sheep can become hefted to one particular local pasture (heft) so they do not roam freely in unfenced landscapes. Lambs learn the heft from ewes and if whole flocks are culled it must be retaught to the replacement animals.

Flock behavior in sheep is generally only exhibited in groups of four or more sheep; fewer sheep may not react as expected when alone or with few other sheep. Being a prey species, the primary defense mechanism of sheep is to flee from danger when their flight zone is entered. Cornered sheep may charge and butt, or threaten by hoof stamping and adopting an aggressive posture. This is particularly true for ewes with newborn lambs.

In regions where sheep have no natural predators, none of the native breeds of sheep exhibit a strong flocking behavior.

## Herding

Farmers exploit flocking behavior to keep sheep together on unfenced pastures such as hill farming, and to move them more easily. For this purpose shepherds may use herding dogs in this effort, with a highly bred herding ability. Sheep are food-oriented, and association of humans with regular feeding often results in sheep soliciting people for food. Those who are moving sheep may exploit this behavior by leading sheep with buckets of feed.

Escaped sheep being led back to pasture with the enticement of food.
This method of moving sheep works best with smaller flocks.

## Dominance Hierarchy

Sheep establish a dominance hierarchy through fighting, threats and competitiveness. Dominant animals are inclined to be more aggressive with other sheep, and usually feed first at troughs. Primarily among rams, horn size is a factor in the flock hierarchy. Rams with different size horns may be less inclined to fight to establish the dominance order, while rams with similarly sized horns are more so. Merinos have an almost linear hierarchy whereas there is a less rigid structure in Border Leicesters when a competitive feeding situation arises.

In sheep, position in a moving flock is highly correlated with social dominance, but there is no definitive study to show consistent voluntary leadership by an individual sheep.

## Intelligence and Learning Ability

Sheep are frequently thought of as unintelligent animals. Their flocking behavior and quickness to flee and panic can make shepherding a difficult endeavor for the uninitiated. Despite these perceptions, a University of Illinois monograph on sheep reported their intelligence to be just below that of pigs and on par with that of cattle. Sheep can recognize individual human and ovine faces and remember them for years. In addition to long-term facial recognition of individuals, sheep can also differentiate emotional states through facial characteristics. If worked with patiently, sheep may learn their names, and many sheep are trained to be led by halter for showing and other purposes. Sheep have also responded well to clicker training. Sheep have been used as pack animals; Tibetan nomads distribute baggage equally throughout a flock as it is herded between living sites.

It has been reported that some sheep have apparently shown problem-solving abilities; a flock in West Yorkshire, England allegedly found a way to get over cattle grids by rolling on their backs, although documentation of this has relied on anecdotal accounts.

## Vocalisations

Sounds made by domestic sheep include bleats, grunts, rumbles and snorts. Bleating ("baaing") is used mostly for contact communication, especially between dam and lambs, but also at times between other flock members. The bleats of individual sheep are distinctive, enabling the ewe and her lambs to recognize each other's vocalizations. Vocal communication between lambs and their dam declines to a very low level within several weeks after parturition. A variety of bleats may be heard, depending on sheep age and circumstances. Apart from contact communication, bleating may signal distress, frustration or impatience; however, sheep are usually silent when in pain. Isolation commonly prompts bleating by sheep. Pregnant ewes may grunt when in labor. Rumbling sounds are made by the ram during courting; somewhat similar rumbling sounds may be made by the ewe, especially when with her neonate lambs. A snort (explosive exhalation through the nostrils) may signal aggression or a warning, and is often elicited from startled sheep.

## Senses

Lamb

In sheep breeds lacking facial wool, the visual field is wide. In 10 sheep (Cambridge, Lleyn and Welsh Mountain breeds, which lack facial wool), the visual field ranged from 298° to 325°, averaging 313.1°, with binocular overlap ranging from 44.5° to 74°, averaging 61.7°. In some breeds, unshorn facial wool can limit the visual field; in some individuals, this may be enough to cause "wool blindness". In 60 Merinos, visual fields ranged from 219.1° to 303.0°, averaging 269.9°, and the binocular field ranged from 8.9° to 77.7°, averaging 47.5°; 36% of the measurements were limited by wool, although photographs of the experiments indicate that only limited facial wool regrowth had occurred since shearing. In addition to facial wool (in some breeds), visual field limitations can include ears and (in some breeds) horns, so the visual field can be extended by tilting the head. Sheep eyes exhibit very low hyperopia and little astigmatism. Such visual characteristics are likely to produce a well-focused retinal image of objects in both the middle and long distance. Because sheep eyes have no accommodation, one might expect the image of very near objects to be blurred, but a rather clear near image could be provided by the tapetum and large retinal image of the sheep's eye, and adequate close vision may occur at muzzle length. Good depth perception, inferred from the sheep's sure-footedness, was confirmed in "visual cliff" experiments; behavioral responses indicating depth perception are seen in lambs at one day old. Sheep are thought to have colour vision, and can distinguish between a variety of colours: black, red, brown, green, yellow and white. Sight is a vital part of sheep communication, and when grazing, they maintain visual contact with each other. Each sheep lifts its head upwards to check the position of other sheep in the flock. This constant monitoring is probably what keeps the sheep in a flock as they move along grazing. Sheep become stressed when isolated; this stress is reduced if they are provided with a mirror, indicating that the sight of other sheep reduces stress.

Taste is the most important sense in sheep, establishing forage preferences, with sweet and sour plants being preferred and bitter plants being more commonly rejected. Touch and sight are also important in relation to specific plant characteristics, such as succulence and growth form.

The ram uses his vomeronasal organ (sometimes called the Jacobson's organ) to sense the pheromones of ewes and detect when they are in estrus. The ewe uses her vomeronasal organ for early recognition of her neonate lamb.

## Reproduction

The second of twins being born.

Sheep follow a similar reproductive strategy to other herd animals. A group of ewes is generally mated by a single ram, who has either been chosen by a breeder or (in feral populations) has established dominance through physical contest with other rams. Most sheep are seasonal breeders, although some are able to breed year-round. Ewes generally reach sexual maturity at six to eight months old, and rams generally at four to six months. However, there are exceptions. For example, Finnsheep ewe lambs may reach puberty as early as 3 to 4 months, and Merino ewes sometimes reach puberty at 18 to 20 months. Ewes have estrus cycles about every 17 days, during which they emit a scent and indicate readiness through physical displays towards rams. A minority of rams (8% on average) display a preference for homosexuality and a small number of the females that were accompanied by a male fetus *in utero* are freemartins (female animals that are behaviorally masculine and lack functioning ovaries).

In feral sheep, rams may fight during the rut to determine which individuals may mate with ewes. Rams, especially unfamiliar ones, will also fight outside the breeding period to establish dominance; rams can kill one another if allowed to mix freely. During the rut, even usually friendly rams may become aggressive towards humans due to increases in their hormone levels.

After mating, sheep have a gestation period of about five months, and normal labor takes one to three hours. Although some breeds regularly throw larger litters of lambs, most produce single or twin lambs. During or soon after labor, ewes and lambs may be confined to small lambing jugs, small pens designed to aid both careful observation of ewes and to cement the bond between them and their lambs.

A lamb's first steps.

Ovine obstetrics can be problematic. By selectively breeding ewes that produce multiple offspring with higher birth weights for generations, sheep producers have inadvertently caused some domestic sheep to have difficulty lambing; balancing ease of lambing with high productivity is one of the dilemmas of sheep breeding. In the case of any such problems, those present at lambing may assist the ewe by extracting or repositioning lambs. After the birth, ewes ideally break the amniotic sac (if it is not broken during labor), and begin licking clean the lamb. Most lambs will begin standing within an hour of birth. In normal situations, lambs nurse after standing, receiving vital colostrum milk. Lambs that either fail to nurse or are rejected by the ewe require help to survive, such as bottle-feeding or fostering by another ewe.

## Farming Practices

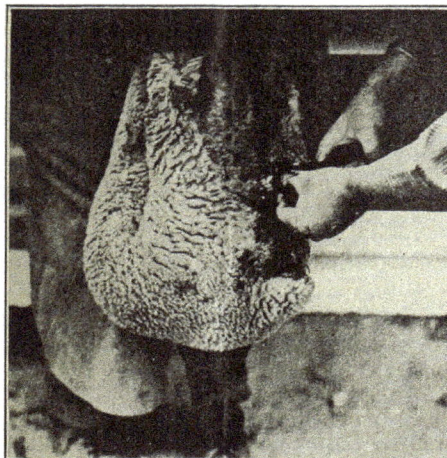
Castrating a docked lamb.

Most lambs begin life being born outdoors. 'Winter lambing' is the practice of impregnating sheep so that they give birth in winter months, meaning their lambs are weaned in spring when pastures are most fertile. While this allows the lambs to grow more quickly, it results in roughly one in every four newborn lambs dying within within a few days of birth due to malnutrition, disease, or exposure to the harsh cold. In the UK, it results in around 4 million newborn lamb deaths, and in Australia the practice leads to 10 to 15 million newborn lamb deaths. However, this is still preferable to the higher feed costs of lambing in warmer months.

After lambs are several weeks old, lamb marking (ear tagging, docking, mulesing, and castrating) is carried out. Vaccinations are usually carried out at this point as well. Ear tags with numbers are attached, or ear marks are applied, for ease of later identification of sheep. The Merino breed, accounting for around 80% of the wool produced in Australia, have been selectively bred to have wrinkled skin resulting in excessive amounts of wool while making them much more prone to fly-strike. To reduce the risk of flystrike caused by soiling for the lambs who make it to summer, their tails are docked, and Merino lambs are often mulesed at the same time, which involves cutting off the skin around their buttocks and the base of their tail with metal shears. If the lambs are younger than 6 months, it is legal to do this in Australia without any pain relief. Tail docking is commonly done for welfare, having been shown to reduce risk of flystrike when compared to the alternative of letting sheep collect waste around their buttocks. Around this time, male lambs are typically cas-trated. Castration is performed on ram lambs not intended for breeding, although some shepherds choose to omit this for ethical, economic or practical reasons. A common castration technique is 'elastration,' which involves a thick rubber band being placed around the base of the infant's scrotum, obstructing the blood supply and causing atrophy. This method causes severe pain to the lambs who are provided no pain relief during the process. Elastration is also commonly used for docking.

10,000 sheep and 200 cattle were sold at this auction in 1916. Heat stress, dehydration, exhaustion, or pre-existing conditions are common causes of deaths at auctions such as this.

Docking and castration are commonly done after 24 hours (to avoid interference with maternal bond-ing and consumption of colostrum) and are often done not later than one week after birth, to minimize pain, stress, recovery time and complications. The first course of vaccinations (commonly anti-clostrid-ial) is commonly given at an age of about 10 to 12 weeks; i.e. when the concentration of maternal anti-bodies passively acquired via colostrum is expected to have fallen low enough to permit development of active immunity. Ewes are often revaccinated annually about 3 weeks before lambing, to provide high antibody concentrations in colostrum during the first several hours after lambing. Ram lambs that will either be slaughtered or separated from ewes before sexual maturity are not usually castrated. Objections to all these procedures have been raised by animal rights groups, but farmers defend them by saying they save money, and inflict only temporary pain.

Sheep do not require expensive housing, such as that used in the intensive farming of chickens or pigs. They are an efficient use of land; roughly six sheep can be kept on the amount that would suffice for a single cow or horse. Sheep can also consume plants, such as noxious weeds, that most other animals will not touch, and produce more young at a faster rate. Also, in contrast to most livestock species, the cost of raising sheep is not necessarily tied to the price of feed crops such as grain, soybeans and corn. Sheep shearers are paid by the number of sheep shorn, not by the hour, making speed often prioritised over precision, and there is no requirement for formal training or accreditation.

Sheep in a slaughterhouse.

After a few years, when they can no longer produce enough wool to be considered profitable, sheep are sent to slaughter and sold as mutton, while lambs raised for meat are killed between 4 and 12 months of age, short of a natural lifespan of 12-14 years. 19 million of the 32 million sheep killed each year in Australia go through saleyards, an intermediary between farms and slaughterhouses or private buyers, where animals also including cattle, calves, horses, poultry and pigs, are auctioned off. Heat stress, dehydration, exhaustion, or pre-existing conditions are common causes of deaths at saleyards.

Around 1.4 million sheep and goats are killed without being stunned each year in the UK using halal practices. Many people in the UK oppose this form of slaughter, yet purchase halal meat unknowingly, since it is sold in most major outlets, including supermarkets and takeaways, without always being labelled as halal. In stun slaughterhouses across the UK, sheep are most commonly stunned using electrical prongs to render them unconscious - however it is estimated that as many as 4 million sheep may be conscious each year whilst they are having their throats slit. Electrical stunning proves regularly ineffective, causing only pain and terrifying the animals even further in their final moments.

## Health

Sheep may fall victim to poisons, infectious diseases, and physical injuries. As a prey species, a sheep's system is adapted to hide the obvious signs of illness, to prevent being targeted by predators. However, some signs of ill health are obvious, with sick sheep eating little, vocalizing excessively, and being generally listless. Throughout history, much of the money and labor of sheep husbandry has aimed to prevent sheep ailments. Historically, shepherds often created remedies by experimentation on the farm. In some developed countries, including the United States, sheep lack the economic importance for drug companies to perform expensive clinical trials required to approve more than a relatively limited number of drugs for ovine use. However, extra-label drug use in sheep production is permitted in many jurisdictions, subject to certain restrictions. In the US, for example, regulations governing extra-label drug use in animals are found in 21 CFR (Code of

Federal Regulations) Part 530. In the 20th and 21st centuries, a minority of sheep owners have turned to alternative treatments such as homeopathy, herbalism and even traditional Chinese medicine to treat sheep veterinary problems. Despite some favorable anecdotal evidence, the effectiveness of alternative veterinary medicine has been met with skepticism in scientific journals. The need for traditional anti-parasite drugs and antibiotics is widespread, and is the main impediment to certified organic farming with sheep.

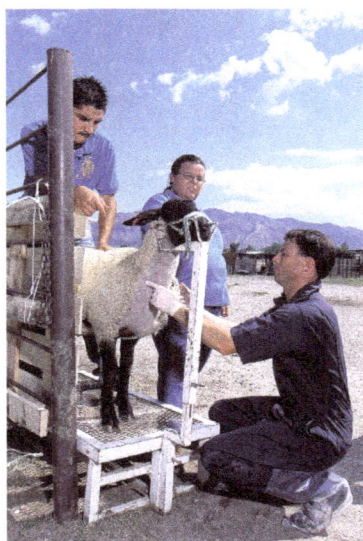

A veterinarian draws blood to test for resistance to scrapie.

Many breeders take a variety of preventive measures to ward off problems. The first is to ensure all sheep are healthy when purchased. Many buyers avoid outlets known to be clearing houses for animals culled from healthy flocks as either sick or simply inferior. This can also mean maintaining a closed flock, and quarantining new sheep for a month. Two fundamental preventive programs are maintaining good nutrition and reducing stress in the sheep. Restraint, isolation, loud noises, novel situations, pain, heat, extreme cold, fatigue and other stressors can lead to secretion of cortisol, a stress hormone, in amounts that may indicate welfare problems. Excessive stress can compromise the immune system. "Shipping fever" (pneumonic mannheimiosis, formerly called pasteurellosis) is a disease of particular concern, that can occur as a result of stress, notably during transport and (or) handling. Pain, fear and several other stressors can cause secretion of epinephrine (adrenaline). Considerable epinephrine secretion in the final days before slaughter can adversely affect meat quality (by causing glycogenolysis, removing the substrate for normal post-slaughter acidification of meat) and result in meat becoming more susceptible to colonization by spoilage bacteria. Because of such issues, low-stress handling is essential in sheep management. Avoiding poisoning is also important; common poisons are pesticide sprays, inorganic fertilizer, motor oil, as well as radiator coolant containing ethylene glycol.

Common forms of preventive medication for sheep are vaccinations and treatments for parasites. Both external and internal parasites are the most prevalent malady in sheep, and are either fatal, or reduce the productivity of flocks. Worms are the most common internal parasites. They are ingested during grazing, incubate within the sheep, and are expelled through the digestive system (beginning the cycle again). Oral anti-parasitic medicines, known as drenches, are given to a flock to treat worms, sometimes after worm eggs in the feces has been counted to assess infestation levels. Afterwards, sheep may be moved to a new pasture to avoid ingesting the same parasites. External sheep parasites include: lice (for different parts of the body), sheep keds, nose bots, sheep itch mites, and maggots. Keds are blood-sucking parasites that cause general malnutrition and decreased productivity, but are not fatal. Maggots are those of the bot fly and the blow-fly. Fly maggots cause the extremely destructive condition of flystrike. Flies lay their eggs in wounds or wet, manure-soiled wool; when the maggots hatch they burrow into a sheep's flesh, eventually causing death if untreated. In addition to other treatments, crutching (shearing wool from a sheep's rump) is a common preventive method. Some countries allow mulesing, a practice that involves stripping away the skin on the rump to prevent fly-strike, normally performed when the sheep is a lamb. Nose bots are fly larvae that inhabit a sheep's sinuses, causing breathing difficulties and discomfort. Common signs are a discharge from the nasal passage, sneezing, and frantic movement such as head shaking. External parasites may be controlled through the use of backliners, sprays or immersive sheep dips.

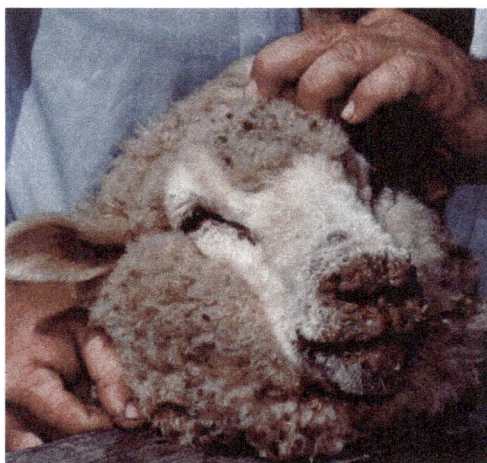

A sheep infected with orf, a disease transmittable to humans through skin contact.

A wide array of bacterial and viral diseases affect sheep. Diseases of the hoof, such as foot rot and foot scald may occur, and are treated with footbaths and other remedies. Foot rot is present in over 97% of flocks in the UK. These painful conditions cause lameness and hinder feeding. Ovine Johne's disease is a wasting disease that affects young sheep. Bluetongue disease is an insect-borne illness causing fever and inflammation of the mucous membranes. Ovine rinderpest (or *peste des petits ruminants*) is a highly contagious and often fatal viral disease affecting sheep and goats. Sheep may also be affected by primary or secondary photosensitization.

A few sheep conditions are transmissible to humans. Orf (also known as scabby mouth, contagious ecthyma or soremouth) is a skin disease leaving lesions that is transmitted through skin-to-skin contact. Cutaneous anthrax is also called woolsorter's disease, as the spores can be transmitted in unwashed wool. More seriously, the organisms that can cause spontaneous enzootic abortion in sheep are easily transmitted to pregnant women. Also of concern are the prion disease scrapie and the virus that causes foot-and-mouth disease (FMD), as both can devastate flocks. The latter poses a slight risk to humans. During the 2001 FMD pandemic in the UK, hundreds of sheep were culled and some rare British breeds were at risk of extinction due to this.

Of the 600,300 sheep lost to the US economy in 2004, 37.3% were lost to predators, while 26.5% were lost to some form of disease. Poisoning accounted for 1.7% of non-productive deaths.

## Predators

Other than parasites and disease, predation is a threat to sheep and the profitability of sheep raising. Sheep have little ability to defend themselves, compared with other species kept as livestock. Even if sheep survive an attack, they may die from their injuries or simply from panic. However, the impact of predation varies dramatically with region. In Africa, Australia, the Americas, and parts of Europe and Asia predators are a serious problem. In the United States, for instance, over one third of sheep deaths in 2004 were caused by predation. In contrast, other nations are virtually devoid of sheep predators, particularly islands known for extensive sheep husbandry. Worldwide, canids—including the domestic dog—are responsible for most sheep deaths. Other animals that occasionally prey on sheep include: felines, bears, birds of prey, ravens and feral hogs.

A lamb being attacked by coyotes with a bite to the throat.

Sheep producers have used a wide variety of measures to combat predation. Pre-modern shepherds used their own presence, livestock guardian dogs, and protective structures such as barns and fencing. Fencing (both regular and electric), penning sheep at night and lambing indoors all continue to be widely used. More modern shepherds used guns, traps, and poisons to kill predators, causing significant decreases in predator populations. In the wake of the environmental and conservation movements, the use of these methods now usually falls under the purview of specially designated government agencies in most developed countries.

The 1970s saw a resurgence in the use of livestock guardian dogs and the development of new methods of predator control by sheep producers, many of them non-lethal. Donkeys and guard llamas have been used since the 1980s in sheep operations, using the same basic principle as livestock guardian dogs. Interspecific pasturing, usually with larger livestock such as cattle or horses, may help to deter predators, even if such species do not actively guard sheep. In addition to animal guardians, contemporary sheep operations may use non-lethal predator deterrents such as motion-activated lights and noisy alarms.

## Economic Importance

Sheep are an important part of the global agricultural economy. However, their once vital status has been largely replaced by other livestock species, especially the pig, chicken, and cow. China,

Australia, India, and Iran have the largest modern flocks, and serve both local and exportation needs for wool and mutton. Other countries such as New Zealand have smaller flocks but retain a large international economic impact due to their export of sheep products. Sheep also play a major role in many local economies, which may be niche markets focused on organic or sustainable agriculture and local food customers. Especially in developing countries, such flocks may be a part of subsistence agriculture rather than a system of trade. Sheep themselves may be a medium of trade in barter economies.

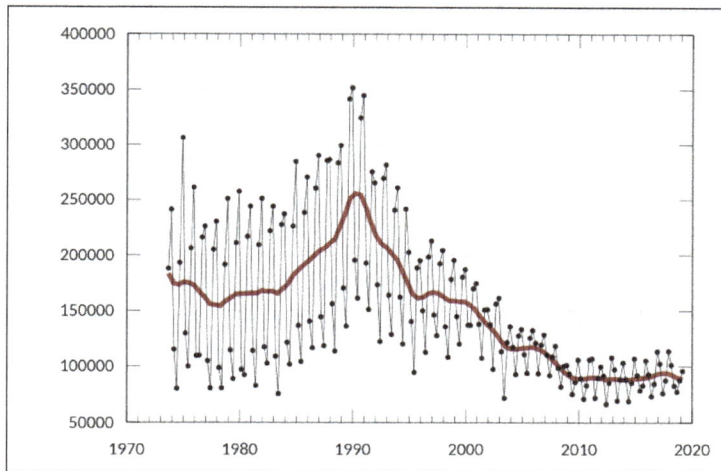

Wool supplied by Australian farmers to dealers (tonnes/quarter) has been in decline since 1990.

Domestic sheep provide a wide array of raw materials. Wool was one of the first textiles, although in the late 20th century wool prices began to fall dramatically as the result of the popularity and cheap prices for synthetic fabrics. For many sheep owners, the cost of shearing is greater than the possible profit from the fleece, making subsisting on wool production alone practically impossible without farm subsidies. Fleeces are used as material in making alternative products such as wool insulation. In the 21st century, the sale of meat is the most profitable enterprise in the sheep industry, even though far less sheep meat is consumed than chicken, pork or beef.

Sheepskin is likewise used for making clothes, footwear, rugs and other products. Byproducts from the slaughter of sheep are also of value: sheep tallow can be used in candle and soap making, sheep bone and cartilage has been used to furnish carved items such as dice and buttons as well as rendered glue and gelatin. Sheep intestine can be formed into sausage casings, and lamb intestine has been formed into surgical sutures, as well as strings for musical instruments and tennis rackets. Sheep droppings, which are high in cellulose, have even been sterilized and mixed with traditional pulp materials to make paper. Of all sheep byproducts, perhaps the most valuable is lanolin: the waterproof, fatty substance found naturally in sheep's wool and used as a base for innumerable cosmetics and other products.

Some farmers who keep sheep also make a profit from live sheep. Providing lambs for youth programs such as 4-H and competition at agricultural shows is often a dependable avenue for the sale of sheep. Farmers may also choose to focus on a particular breed of sheep in order to sell registered purebred animals, as well as provide a ram rental service for breeding. A new option for deriving profit from live sheep is the rental of flocks for grazing; these "mowing services" are hired in order to keep unwanted vegetation down in public spaces and to lessen fire hazard.

Despite the falling demand and price for sheep products in many markets, sheep have distinct economic advantages when compared with other livestock. They do not require expensive housing, such as that used in the intensive farming of chickens or pigs. They are an efficient use of land; roughly six sheep can be kept on the amount that would suffice for a single cow or horse. Sheep can also consume plants, such as noxious weeds, that most other animals will not touch, and produce more young at a faster rate. Also, in contrast to most livestock species, the cost of raising sheep is not necessarily tied to the price of feed crops such as grain, soybeans and corn. Combined with the lower cost of quality sheep, all these factors combine to equal a lower overhead for sheep producers, thus entailing a higher profitability potential for the small farmer. Sheep are especially beneficial for independent producers, including family farms with limited resources, as the sheep industry is one of the few types of animal agriculture that has not been vertically integrated by agribusiness.

## As food

Sheep meat and milk were one of the earliest staple proteins consumed by human civilization after the transition from hunting and gathering to agriculture. Sheep meat prepared for food is known as either mutton or lamb. "Mutton" is derived from the Old French *moton*, which was the word for sheep used by the Anglo-Norman rulers of much of the British Isles in the Middle Ages. This became the name for sheep meat in English, while the Old English word *sceap* was kept for the live animal. Throughout modern history, "mutton" has been limited to the meat of mature sheep usually at least two years of age; "lamb" is used for that of immature sheep less than a year.

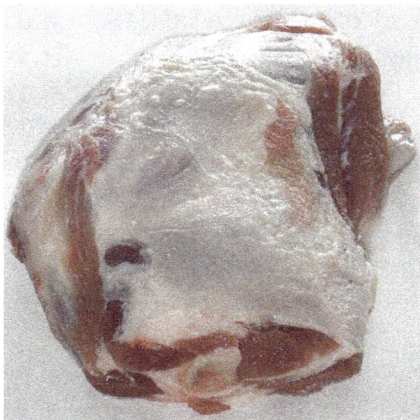

Shoulder of lamb.

In the 21st century, the nations with the highest consumption of sheep meat are the Arab States of the Persian Gulf, New Zealand, Australia, Greece, Uruguay, the United Kingdom and Ireland. These countries eat 14–40 lbs (3–18 kg) of sheep meat per capita, per annum. Sheep meat is also popular in France, Africa (especially the Arab World), the Caribbean, the rest of the Middle East, India, and parts of China. This often reflects a history of sheep production. In these countries in particular, dishes comprising alternative cuts and offal may be popular or traditional. Sheep testicles—called animelles or lamb fries—are considered a delicacy in many parts of the world. Perhaps the most unusual dish of sheep meat is the Scottish haggis, composed of various sheep innards cooked along with oatmeal and chopped onions inside its stomach. In comparison, countries such as the U.S. consume only a pound or less (under 0.5 kg), with Americans eating 50 pounds (22 kg) of pork and 65 pounds (29 kg) of beef. In addition, such countries rarely eat mutton, and may favor the more expensive cuts of lamb: mostly lamb chops and leg of lamb.

Though sheep's milk may be drunk rarely in fresh form, today it is used predominantly in cheese and yogurt making. Sheep have only two teats, and produce a far smaller volume of milk than cows. However, as sheep's milk contains far more fat, solids, and minerals than cow's milk, it is ideal for the cheese-making process. It also resists contamination during cooling better because of its much higher calcium content. Well-known cheeses made from sheep milk include the Feta of Bulgaria and Greece, Roquefort of France, Manchego from Spain, the Pecorino Romano (the Italian word for sheep is *pecore*) and Ricotta of Italy. Yogurts, especially some forms of strained yogurt, may also be made from sheep milk. Many of these products are now often made with cow's milk, especially when produced outside their country of origin. Sheep milk contains 4.8% lactose, which may affect those who are intolerant.

As with other domestic animals, the meat of uncastrated males is inferior in quality, especially as they grow. A "bucky" lamb is a lamb which was not castrated early enough, or which was castrated improperly (resulting in one testicle being retained). These lambs are worth less at market.

## In Science

Sheep are generally too large and reproduce too slowly to make ideal research subjects, and thus are not a common model organism. They have, however, played an influential role in some fields of science. In particular, the Roslin Institute of Edinburgh, Scotland used sheep for genetics research that produced groundbreaking results. In 1995, two ewes named Megan and Morag were the first mammals cloned from differentiated cells. A year later, a Finnish Dorset sheep named Dolly, dubbed "the world's most famous sheep" in *Scientific American*, was the first mammal to be cloned from an adult somatic cell. Following this, Polly and Molly were the first mammals to be simultaneously cloned and transgenic.

A cloned ewe named Dolly was a scientific landmark.

As of 2008, the sheep genome has not been fully sequenced, although a detailed genetic map has been published, and a draft version of the complete genome produced by assembling sheep DNA sequences using information given by the genomes of other mammals. In 2012, a transgenic sheep named "Peng Peng" was cloned by Chinese scientists, who spliced his genes with that of a roundworm (C. elegans) in order to increase production of fats healthier for human consumption.

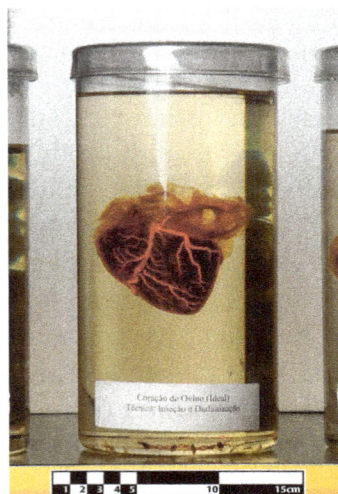
Ovine heart.

In the study of natural selection, the population of Soay sheep that remain on the island of Hirta have been used to explore the relation of body size and coloration to reproductive success. Soay sheep come in several colors, and researchers investigated why the larger, darker sheep were in decline; this occurrence contradicted the rule of thumb that larger members of a population tend to be more successful reproductively. The feral Soays on Hirta are especially useful subjects because they are isolated.

Sheep are one of the few animals where the molecular basis of the diversity of male sexual preferences has been examined. However, this research has been controversial, and much publicity has been produced by a study at the Oregon Health and Science University that investigated the mechanisms that produce homosexuality in rams. Organizations such as PETA campaigned against the study, accusing scientists of trying to cure homosexuality in the sheep. OHSU and the involved scientists vehemently denied such accusations.

Domestic sheep are sometimes used in medical research, particularly for researching cardiovascular physiology, in areas such as hypertension and heart failure. Pregnant sheep are also a useful model for human pregnancy, and have been used to investigate the effects on fetal development of malnutrition and hypoxia. In behavioral sciences, sheep have been used in isolated cases for the study of facial recognition, as their mental process of recognition is qualitatively similar to humans.

## Domestic Pig

The domestic pig (Sus scrofa domesticus or only Sus domesticus), often called swine, hog, or simply pig when there is no need to distinguish it from other pigs, is a domesticated large, even-toed ungulate. It is variously considered a subspecies of the Eurasian boar or a distinct species. The domestic pig's head-plus-body-length ranges from 0.9 to 1.8 m (35 to 71 in), and adult pigs typically weigh between 50 and 350 kg (110 and 770 lb), with well-fed individuals often exceeding this weight range. The size and weight of a hog largely depends on its breed. Compared to other artiodactyls, its head is relatively long, pointed, and free of warts. Even-toed ungulates are generally herbivorous, but the domestic pig is an omnivore, like its wild relative.

When used as livestock, domestic pigs are farmed primarily for the consumption of their flesh, called pork. The animal's bones, hide, and bristles are also used in commercial products. Domestic pigs, especially miniature breeds, are kept as pets.

## Biology

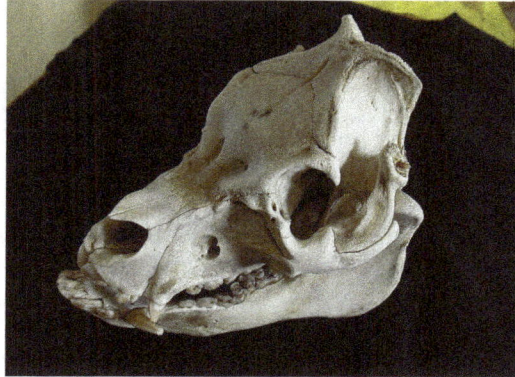

Domestic pig skull.

The domestic pig typically has a large head, with a long snout which is strengthened by a special prenasal bone and a disk of cartilage at the tip. The snout is used to dig into the soil to find food, and is a very acute sense organ. The dental formula of adult pigs is 3.1.4.33.1.4.3, giving a total of 44 teeth. The rear teeth are adapted for crushing. In the male the canine teeth can form tusks, which grow continuously and are sharpened by constantly being ground against each other.

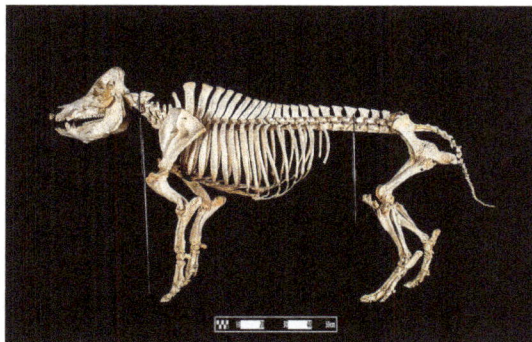

Skeleton specimen of a swine.

Four hoofed toes are on each foot, with the two larger central toes bearing most of the weight, but the outer two also being used in soft ground.

Most domestic pigs have rather a bristled sparse hair covering on their skin, although woolly-coated breeds such as the Mangalitsa exist.

Pigs possess both apocrine and eccrine sweat glands, although the latter appear limited to the snout and dorsonasal areas. Pigs, however, like other "hairless" mammals (e.g. elephants, rhinos, and mole-rats), do not use thermal sweat glands in cooling. Pigs are also less able than many other mammals to dissipate heat from wet mucous membranes in the mouth through panting. Their thermoneutral zone is 16 to 22 °C (61 to 72 °F). At higher temperatures, pigs lose heat by wallowing in mud or water via evaporative cooling; although it has been suggested that wallowing may serve other functions, such as protection from sunburn, ecto-parasite control, and scent-marking.

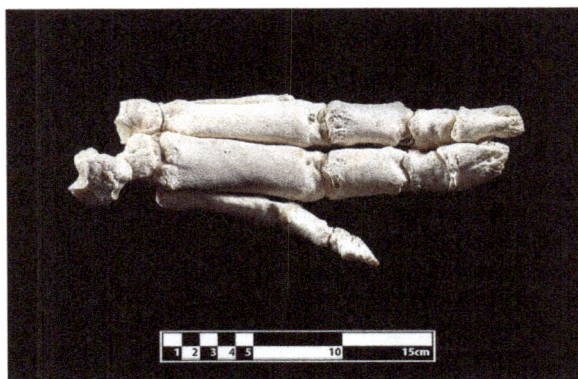

Bones of a domestic pig's foot.

Pigs are one of four known mammalian species which possess mutations in the nicotinic acetylcholine receptor that protect against snake venom. Mongooses, honey badgers, hedgehogs, and pigs all have modifications to the receptor pocket which prevents the snake venom α-neurotoxin from binding. These represent four separate, independent mutations.

Domestic pigs have small lungs in relation to their body size, and are thus more susceptible than other domesticated animals to fatal bronchitis and pneumonia.

## Taxonomy

The domestic pig is most often considered to be a subspecies of the wild boar, which was given the name *Sus scrofa* by Carl Linnaeus in 1758; following from this, the formal name of the domestic pig is *Sus scrofa domesticus*. However, in 1777, Johann Christian Polycarp Erxleben classified the domestic pig as a separate species from the wild boar. He gave it the name *Sus domesticus*, which is still used by some taxonomists.

## Reproduction

Female pigs reach sexual maturity at 3–12 months of age, and come into estrus every 18–24 days if they are not successfully bred. The variation in ovulation rate can be attributed to intrinsic factors such as age and genotype, as well as extrinsic factors like nutrition, environment and the supplementation of exogenous hormones. The gestation period averages 112–120 days. Estrus lasts two to three days, and the female's displayed receptiveness to mate is known as standing heat. Standing heat is a reflexive response that is stimulated when the female is in contact with the saliva of a sexually mature boar. Androstenol is one of the pheromones produced in the submaxillary salivary glands of boars that will trigger the female's response. The female cervix contains a series of five interdigitating pads, or folds, that will hold the boar's corkscrew-shaped penis during copulation. Females have bicornuate uteruses and two conceptuses must be present in both uterine horns for pregnancy to be established. Maternal recognition of pregnancy in pigs occurs on days 11 to 12 of pregnancy and is marked by progesterone production from a functioning corpus luteum (CL). To avoid luteolysis by PGF2α, rescuing of the CL must occur via embryonic signaling of estradiol 17β and PGE2. This signaling acts on both the endometrium and luteal tissue to prevent the regression of the CL by activation of genes that are responsible for CL maintenance. During mid to late pregnancy, the CL relies primarily on luteinizing hormone (LH) for maintenance until parturition. Animal nutrition is important prior to reproduction and during gestation to ensure optimum reproductive performance is achieved.

## Behavior

In many ways, their behavior appears to be intermediate between that of other artiodactyls and of carnivores. Domestic pigs seek out the company of other pigs, and often huddle to maintain physical contact, although they do not naturally form large herds. They typically live in groups of about 8-10 adult sows, some young individuals, and some single males.

Domestic pigs in a wallow.

Because of their relative lack of sweat glands, pigs often control their body temperature using behavioral thermoregulation. Wallowing, which often consists of coating the body with mud, is a behavior frequently exhibited by pigs. They do not submerge completely under the mud, but vary the depth and duration of wallowing depending on environmental conditions. Typically, adult pigs start wallowing once the ambient temperature is around 17-21 °C. They cover themselves from head to toe in mud. Pigs may use mud as a sunscreen, or as a method of keeping parasites away. Most bristled pigs will "blow their coat", meaning that they shed most of the longer, more-coarse stiff hair once a year, usually in spring or early summer, to prepare for the warmer months ahead.

If conditions permit, domestic pigs feed continuously for many hours and then sleep for many hours, in contrast to ruminants which tend to feed for a short time and then sleep for a short time. Pigs are omnivorous, and are highly versatile in their feeding behavior. As they are foraging animals, they primarily eat leaves, stems, roots, fruits, and flowers. Domestic pigs are highly intelligent animals, on par with dogs, and according to David DiSalvo's writing in *Forbes*, they are "widely considered the smartest domesticated animal in the world. Pigs can move a cursor on a video screen with their snouts and even learn to distinguish between the scribbles they knew from those they saw for the first time."

## Rooting

Rooting is an instinctual behavior in pigs that is characterized by a pig nudging its snout into something. Similar to a cat's kneading, rooting is found comforting, first happens when they are born in order to obtain their mother's milk, and can become a habitual, obsessive behavior which is most prominent in animals weaned too early. Often, pigs will root and dig into the ground to forage for food. Rooting is known to also be used as a means of communication. A nose ring is sometimes put through the septum: it discourages rooting because that is painful.

The breed known as the kunekune hardly ever roots, as it sustains itself feeding on nothing other than grass leaves. Not having to root around in the soil to find underground food (e.g. tubers), it thus has evolved to, for the most part, not possess these instincts.

## Nest-building

A behavioral characteristic of domestic pigs which they share with carnivores is nest-building. Sows root in the ground to create depressions and then build nests in which to give birth. First, the sow digs a depression about the size of her body. She then collects twigs and leaves, and carries these in her mouth to the depression, building them into a mound. She distributes the softer, finer material to the centre of the mound using her feet. When the mound reaches the desired height, she places large branches, up to 2 metres in length, on the surface. She enters into the mound and roots around to create a depression within the gathered material. She then gives birth in a lying position, which, again, is different from other artiodactyls, which usually give birth in a standing position.

Nest-building behavior is an important part in the process of pre and post-partum maternal behavior. Nest-building will occur during the last 24 hours before the onset of farrowing, and becomes most intense during 12 to 6 hours before farrowing. Nest-building is divided into two phases: one of which is the initial phase of rooting in ground while the second phase is the collecting, carrying and arranging of the nest material. The sow will separate from the group and seek a suitable nest site with some shelter from rain and wind that has well-drained soil. This nest-building behavior is performed to provide the offspring with shelter, comfort, and thermoregulation. The nest will provide protection against weather and predators, while keeping the piglets close to the sow and away from the rest of the herd. This ensures they do not get trampled on and that other piglets are not stealing milk from the sow. Nest-building can be influenced by internal and external stimuli. Internal hormonal changes and the completion of one nesting phase are indicators of this maternal behavior. The onset is triggered by the rise in prolactin levels, which is caused by a decrease in progesterone and an increase in prostaglandin, while the gathering of the nest material seems to be regulated more by external stimuli such as temperature. The longer time spent on nest-building will increase pre-partum oxytocin.

## Nursing and Suckling Behavior

Compared to most other mammals, pigs display complex nursing and suckling behavior. Nursing occurs every 50–60 minutes, and the sow requires stimulation from piglets before milk let-down. Sensory inputs (vocalisation, odours from mammary and birth fluids and hair patterns of the sow) are particularly important immediately post-birth to facilitate teat location by the piglets. Initially, the piglets compete for position at the udder, then each piglet massages around its respective teat with its snout, during which time the sow grunts at slow, regular intervals. Each series of grunts varies in frequency, tone and magnitude, indicating the stages of nursing to the piglets.

Sow with prominent nipples. Pigs typically have 12–14 nipples.

The phase of competition for teats and of nosing the udder lasts for about one minute, and ends when milk flow begins. In the third phase, the piglets hold the teats in their mouths and suck with slow mouth movements (one per second), and the rate of the sow's grunting increases for approximately 20 seconds. The grunt peak in the third phase of suckling does not coincide with milk ejection, but rather the release of oxytocin from the pituitary into the bloodstream. Phase four coincides with the period of main milk flow (10–20 seconds) when the piglets suddenly withdraw slightly from the udder and start sucking with rapid mouth movements of about three per second. The sow grunts rapidly, lower in tone and often in quick runs of three or four, during this phase. Finally, the flow stops and so does the grunting of the sow. The piglets may then dart from teat to teat and recommence suckling with slow movements, or nosing the udder. Piglets massage and suckle the sow's teats after milk flow ceases as a way of letting the sow know their nutritional status. This helps her to regulate the amount of milk released from that teat in future sucklings. The more intense the post-feed massaging of a teat, the greater the future milk release from that teat will be.

## Teat Order

A sow with suckling piglets.

In pigs, dominance hierarchies can be formed at a very early age. Domestic piglets are highly precocious and within minutes of being born, or sometimes seconds, will attempt to suckle. The piglets are born with sharp teeth and fight to develop a teat order as the anterior teats produce a greater quantity of milk. Once established, this teat order remains stable with each piglet tending to feed from a particular teat or group of teats. Stimulation of the anterior teats appears to be important in causing milk letdown, so it might be advantageous to the entire litter to have these teats occupied by healthy piglets. Using an artificial sow to rear groups of piglets, recognition of a teat in a particular area of the udder depended initially on visual orientation by means of reference points on the udder to find the area, and then the olfactory sense for the more accurate search within that area.

## Senses

Pigs have panoramic vision of approximately 310° and binocular vision of 35° to 50°. It is thought they have no eye accommodation. Other animals that have no accommodation, e.g. sheep, lift their heads to see distant objects. The extent to which pigs have colour vision is still a source of some debate; however, the presence of cone cells in the retina with two distinct wavelength sensitivities (blue and green) suggests that at least some colour vision is present.

Pigs have a well-developed sense of smell, and use is made of this in Europe where they are trained to locate underground truffles. Olfactory rather than visual stimuli are used in the identification of other pigs. Hearing is also well developed, and localisation of sounds is made by moving the head. Pigs use auditory stimuli extensively as a means of communication in all social activities. Alarm or aversive stimuli are transmitted to other pigs not only by auditory cues but also by pheromones. Similarly, recognition between the sow and her piglets is by olfactory and vocal cues.

## Breeds

Many breeds of domestic pig exist; in many colors, shapes, and sizes. According to The Livestock Conservancy, as of 2016, three breeds of domestic pig are critically rare (having a global population of fewer than 2000). They are the Choctaw hog, the Mulefoot, and the Ossabaw Island pig. The known smallest domestic pig breed in the world is the Göttingen minipig, typically weighing about 26 kilograms (57 lb) as a healthy, full-grown adult.

## In Agriculture

A Large White, a breed commonly used in meat production.

When in use as livestock, the domestic pig is mostly farmed for its meat, pork. Other food products made from pigs include pork sausage (which includes casings that are made from the intestines), bacon, gammon, ham and pork rinds. The head of a pig can be used to make a preserved jelly called head cheese, which is sometimes known as brawn. Liver, chitterlings, blood (for black pudding), and other offal from pigs are also widely used for food. In some religions, such as Judaism and Islam, pork is a taboo food.

The use of pig milk for human consumption does take place, but as there are certain difficulties in obtaining it, there is little commercial production.

Livestock pigs are exhibited at agricultural shows, judged either as stud stock compared to the standard features of each pig breed, or in commercial classes where the animals are judged primarily on their suitability for slaughter to provide premium meat.

The skin of pigs is used to produce seat covers, apparel, pork rinds, and other items.

In some developing and developed nations, the domestic pig is usually raised outdoors in yards or fields. In some areas, pigs are allowed to forage in woods where they may be taken care of by swineherds. In industrialized nations such as the United States, domestic pig farming has switched from the traditional pig farm to large-scale intensive pig farms. This has resulted in lower production costs, but can cause significant cruelty problems. As consumers have become concerned with humane treatment of livestock, demand for pasture-raised pork in these nations has increased.

## As Pets

Vietnamese pot-bellied pigs, a miniature breed of domestic pig, have made popular pets in the United States, beginning in the latter half of the 20th century. These pot-bellied pigs were soon crossbred with a variety of other small breeds, such as the Göttingen minipig, with separate locations breeding different lineages. These crossbred miniature pigs soon gained attention, even more so than the original Vietnamese pot-bellied. As a result of this, many pet pigs are now of unknown genetic descent.

A crossbred miniature pig.

Domestic pigs are highly intelligent, social creatures. They are considered hypoallergenic, and are known to do quite well with people who have usual animal allergies. Since these animals are known to have a life expectancy of 15 to 20 years, they require a long-term commitment.

Another crossbred "Salt & Pepper" miniature pig.

## Care

Male and female swine that have not been de-sexed may express unwanted aggressive behavior, and are prone to developing serious health issues. Regular trimming of the hooves is necessary; hooves left untreated cause major pain in the pig, can create malformations in bone structure, and may cause it to be more susceptible to fungal growth between crevices of the hoof, or between the cracks in a split hoof. Male pigs, especially when left unaltered, can grow large, sharp tusks which may continue growing for years. Domestic owners may wish to keep their pigs' tusks trimmed back, or have them removed entirely.

As prey animals, pigs' natural instinctive behavior causes them to have a strong fear of being picked up, but they will usually calm down once placed back onto the ground. This instinctual fear may be lessened if the pig has been frequently held since infancy. When holding a pig, supporting it under the legs makes being held not as stressful for the animal. Pigs need enrichment activities to keep their intelligent minds occupied; if pigs get bored, they often become destructive. As rooting is found to be comforting, pigs kept in the house may root household objects, furniture or surfaces. While some owners are known to pierce their pigs' noses to discourage rooting behavior, the efficacy and humaneness of this practice is questionable. As such, indoor pigs should be provided with a box with rocks, soil, straw, and/or other material for them to root in instead.

## In Human Medical Applications

The domestic pig, both as a live animal and source of post-mortem tissues, is one of the most valuable animal models used in biomedical research today, because of its biological, physiological and anatomical similarities to human beings. For instance, human skin is very similar to pig skin, therefore pig skin has been used in many preclinical studies. Porcine are used in finding treatments, cures for diseases, xenotransplantation, and for general education. They are also used in the development of medical instruments and devices, surgical techniques and instrumentation, and FDA-approved research. As part of animal conservation (The Three Rs (animals)), these animals contribute to the reduction methods for animal research, as they supply more information from fewer animals used, for a lower cost.

## Xenotransplantation

Pigs are currently thought to be the best non-human candidates for organ donation to humans. The risk of cross-species disease transmission is decreased because of their increased phylogenetic distance from humans. They are readily available, their organs are anatomically comparable in size, and new infectious agents are less likely since they have been in close contact with humans through domestication for many generations.

To date, no xenotransplantation trials have been entirely successful due to obstacles arising from the response of the recipient's immune system—generally more extreme than in allotransplantations, ultimately results in rejection of the xenograft, and in some cases result in the death of the recipient—including hyperacute rejection, acute vascular rejection, cellular rejection and chronic rejection. An early major breakthrough was the 1,3 galactosyl transferase gene knockout.

Examples of viruses carried by pigs include porcine herpesvirus, rotavirus, parvovirus, and circovirus. Of particular concern are PERVs (porcine endogenous retroviruses), vertically transmitted microbes that embed in swine genomes. The risks with xenosis are twofold, as not only could the individual become infected, but a novel infection could initiate an epidemic in the human population. Because of this risk, the FDA has suggested any recipients of xenotransplants shall be closely monitored for the remainder of their life, and quarantined if they show signs of xenosis.

Pig cells have been engineered to inactivate all 62 PERVs in the genome using CRISPR Cas9 genome editing technology, and eliminated infection from the pig to human cells in culture.

## Goats

The domestic goat or simply goat (*Capra aegagrus hircus*) is a subspecies of *C. aegagrus* domesticated from the wild goat of Southwest Asia and Eastern Europe. The goat is a member of the animal family Bovidae and the subfamily Caprinae, meaning it is closely related to the sheep. There are over 300 distinct breeds of goat. Goats are one of the oldest domesticated species of animal, and have been used for milk, meat, fur and skins across much of the world. Milk from goats is often turned into goat cheese.

Female goats are referred to as *does* or *nannies*, intact males are called *bucks* or *billies* and juvenile goats of both sexes are called *kids*. Castrated males are called *wethers*. While the words *hircine* and *caprine* both refer to anything having a goat-like quality, *hircine* is used most often to emphasize the distinct smell of domestic goats.

In 2011, there were more than 924 million goats living in the world, according to the UN Food and Agriculture Organization.

## Anatomy and Health

Each recognized breed of goat has specific weight ranges, which vary from over 140 kg (300 lb) for bucks of larger breeds such as the Boer, to 20 to 27 kg (45 to 60 lb) for smaller goat does. Within each breed, different strains or bloodlines may have different recognized sizes. At the bottom of the size range are miniature breeds such as the African Pygmy, which stand 41 to 58 cm (16 to 23 in) at the shoulder as adults.

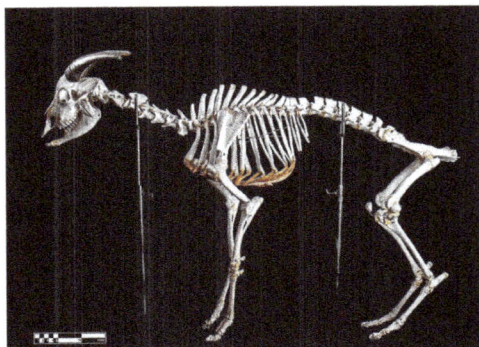
Skeleton (Capra hircus).

Eye with horizontal pupil.

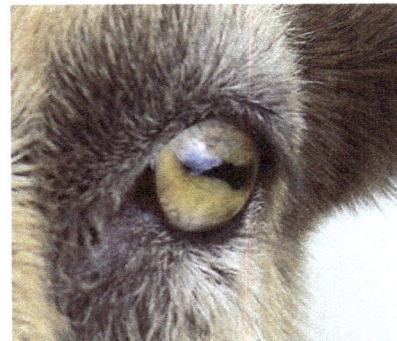
A white Irish goat with horns.

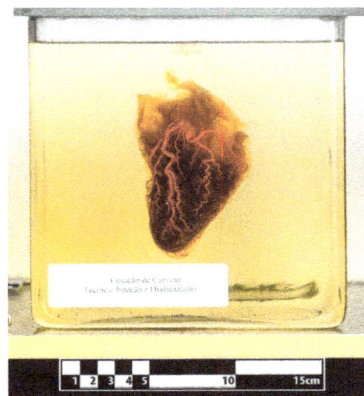
Goat heart. Specimen clarified for visualization of ana- tomical structures.

## Horns

Most goats naturally have two horns, of various shapes and sizes depending on the breed. There have been incidents of polycerate goats (having as many as eight horns), although this is a genetic rarity thought to be inherited. Unlike cattle, goats have not been successfully bred to be reliably polled, as the genes determining sex and those determining horns are closely linked. Breeding together two genetically polled goats results in a high number of intersex individuals among the offspring, which are typically sterile. Their horns are made of living bone surrounded by keratin and other proteins, and are used for defense, dominance, and territoriality.

## Digestion and Lactation

Goats are ruminants. They have a four-chambered stomach consisting of the rumen, the reticulum, the omasum, and the abomasum. As with other mammal ruminants, they are even-toed ungulates. The females have an udder consisting of two teats, in contrast to cattle, which have four teats. An exception to this is the Boer goat, which sometimes may have up to eight teats.

## Eyes

Goats have horizontal, slit-shaped pupils. Because goats' irises are usually pale, their contrasting pupils are much more noticeable than in animals such as cattle, deer, most horses and many sheep, whose similarly horizontal pupils blend into a dark iris and sclera.

## Beards

Both male and female goats have beards, and many types of goat (most commonly dairy goats, dairy-cross Boers, and pygmy goats) may have wattles, one dangling from each side of the neck.

## Tan

Brown/tan Goat with some white spotting.

Goats expressing the tan pattern have coats pigmented completely with phaeomelanin (tan/brown pigment). The allele which codes for this pattern is located at the *agouti locus* of the goat genome. It is completely dominant to all other alleles at this locus. There are multiple modifier genes which control how much tan pigment is actually expressed, so a tan-patterned goat can have a coat ranging from pure white to deep red.

## Reproduction

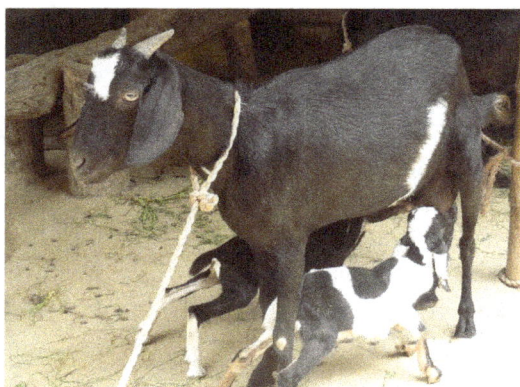
Goat kid.

Goats reach puberty between three and 15 months of age, depending on breed and nutritional status. Many breeders prefer to postpone breeding until the doe has reached 70% of the adult weight. However, this separation is rarely possible in extensively managed, open-range herds.

A two-month-old goat kid in a field of capeweed.

In temperate climates and among the Swiss breeds, the breeding season commences as the day length shortens, and ends in early spring or before. In equatorial regions, goats are able to breed at any time of the year. Successful breeding in these regions depends more on available forage than on day length. Does of any breed or region come into estrus (heat) every 21 days for two to 48 hours. A doe in heat typically flags (vigorously wags) her tail often, stays near the buck if one is present, becomes more vocal, and may also show a decrease in appetite and milk production for the duration of the heat.

A female goat and two kids.

Bucks (intact males) of Swiss and northern breeds come into rut in the fall as with the does' heat cycles. Bucks of equatorial breeds may show seasonal reduced fertility, but as with the does, are capable of breeding at all times. Rut is characterized by a decrease in appetite and obsessive interest in the does. A buck in rut will display flehmen lip curling and will urinate on his forelegs and face. Sebaceous scent glands at the base of the horns add to the male goat's odor, which is important to make him attractive to the female. Some does will not mate with a buck which has been descented.

In addition to natural, traditional mating, artificial insemination has gained popularity among goat breeders, as it allows easy access to a wide variety of bloodlines.

Gestation length is approximately 150 days. Twins are the usual result, with single and triplet births also common. Less frequent are litters of quadruplet, quintuplet, and even sextuplet kids.

Birthing, known as kidding, generally occurs uneventfully. Just before kidding, the doe will have a sunken area around the tail and hip, as well as heavy breathing. She may have a worried look, become restless and display great affection for her keeper. The mother often eats the placenta, which gives her much-needed nutrients, helps stanch her bleeding, and parallels the behavior of wild herbivores, such as deer, to reduce the lure of the birth scent for predators.

Freshening (coming into milk production) occurs at kidding. Milk production varies with the breed, age, quality, and diet of the doe; dairy goats generally produce between 680 and 1,810 kg (1,500 and 4,000 lb) of milk per 305-day lactation. On average, a good quality dairy doe will give at least 3 kg (6 lb) of milk per day while she is in milk. A first-time milker may produce less, or as much as 7 kg (16 lb), or more of milk in exceptional cases. After the lactation, the doe will "dry off", typically after she has been bred. Occasionally, goats that have not been bred and are continuously milked will continue lactation beyond the typical 305 days. Meat, fiber, and pet breeds are not usually milked and simply produce enough for the kids until weaning.

Male lactation is also known to occur in goats.

## Diet

Goats are reputed to be willing to eat almost anything, including tin cans and cardboard boxes. While goats will not actually eat inedible material, they are browsing animals, not grazers like cattle and sheep, and (coupled with their highly curious nature) will chew on and taste just about anything remotely resembling plant matter to decide whether it is good to eat, including cardboard, clothing and paper (such as labels from tin cans). The unusual smells of leftover food in discarded cans or boxes may further stimulate their curiosity.

A domestic goat feeding in a field of capeweed, a weed which is toxic to most stock animals.

Aside from sampling many things, goats are quite particular in what they actually consume, preferring to browse on the tips of woody shrubs and trees, as well as the occasional broad-leaved plant. However, it can fairly be said that their plant diet is extremely varied, and includes some species which are otherwise toxic. They will seldom consume soiled food or contaminated water unless facing starvation. This is one reason goat-rearing is most often free-ranging, since stall-fed goat-rearing involves extensive upkeep and is seldom commercially viable.

Goats prefer to browse on vines, such as kudzu, on shrubbery and on weeds, more like deer than sheep, preferring them to grasses. Nightshade is poisonous; wilted fruit tree leaves can also kill goats. Silage (fermented corn stalks) and haylage (fermented grass hay) can be used if consumed immediately after opening – goats are particularly sensitive to *Listeria* bacteria that can grow in fermented feeds. Alfalfa, a high-protein plant, is widely fed as hay; fescue is the least palatable and least nutritious hay. Mold in a goat's feed can make it sick and possibly kill it.

In various places in China, goats are used in the production of tea. Goats are released onto the tea terraces where they avoid consuming the green tea leaves (which contain bitter tasting substances) but instead eat the weeds. The goats' droppings fertilise the tea plants.

The digestive physiology of a very young kid (like the young of other ruminants) is essentially the same as that of a monogastric animal. Milk digestion begins in the abomasum, the milk having by-passed the rumen via closure of the reticuloesophageal groove during suckling. At birth, the rumen is undeveloped, but as the kid begins to consume solid feed, the rumen soon increases in size and in its capacity to absorb nutrients.

The adult size of a particular goat is a product of its breed (genetic potential) and its diet while growing (nutritional potential). As with all livestock, increased protein diets (10 to 14%) and sufficient calories during the prepuberty period yield higher growth rates and larger eventual size than lower protein rates and limited calories. Large-framed goats, with a greater skeletal size, reach mature weight at a later age (36 to 42 months) than small-framed goats (18 to 24 months) if both are fed to their full potential. Large-framed goats need more calories than small-framed goats for maintenance of daily functions.

## Behavior

Goats are naturally curious. They are also agile and well known for their ability to climb and balance in precarious places. This makes them the only ruminant to regularly climb trees. Due to their agility and inquisitiveness, they are notorious for escaping their pens by testing fences and enclosures, either intentionally or simply because they are used to climbing. If any of the fencing can be overcome, goats will almost inevitably escape. Due to their intelligence, once a goat has discovered a weakness in the fence, they will exploit it repeatedly, and other goats will observe and quickly learn the same method.

Goats establish a dominance hierarchy in flocks, sometimes through head butting.

Glycerinated goat tongue.

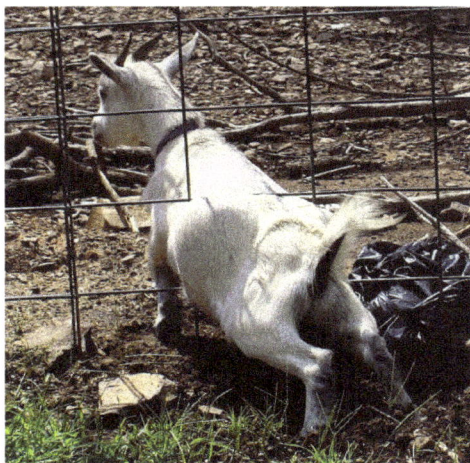

Goats are well known for being hard to contain with fencing.

Goats explore anything new or unfamiliar in their surroundings, primarily with their prehensile upper lip and tongue, by nibbling at them, occasionally even eating them.

When handled as a group, goats tend to display less herding behavior than sheep. When grazing undisturbed, they tend to spread across the field or range, rather than feed side-by-side as do sheep. When nursing young, goats will leave their kids separated ("lying out") rather than clumped, as do sheep. They will generally turn and face an intruder and bucks are more likely to charge or butt at humans than are rams.

A study by Queen Mary University reports that goats try to communicate with people in the same manner as domesticated animals such as dogs and horses. Goats were first domesticated as livestock more than 10,000 years ago. Research conducted to test communication skills found that the goats will look to a human for assistance when faced with a challenge that had previously been mastered, but was then modified. Specifically, when presented with a box, the goat was able to remove the lid and retrieve a treat inside, but when the box was turned so the lid could not be removed, the goat would turn and gaze at the person and move toward them, before looking back toward the box. This is the same type of complex communication observed by animals bred as domestic pets, such as dogs. Researchers believe that better understanding of human-goat interaction could offer overall improvement in the animals' welfare. The field of anthrozoology has established that domesticated animals have the capacity for complex communication with humans when in 2015 a Japanese scientist determined that levels of oxytocin did increase in human subjects when dogs were exposed to a dose of the "love hormone", proving that a human-animal bond does exist. This is the same affinity that was proven with the London study above; goats are intelligent, capable of complex communication, and able to form bonds. Despite having the reputation of being slightly rebellious, more and more people today are choosing more exotic companion animals like goats. Goats are herd animals and typically prefer the company of other goats, but because of their herd mentality, they will follow their owners around just the same.

## Diseases

While goats are generally considered hardy animals and in many situations receive little medical care, they are subject to a number of diseases. Among the conditions affecting goats are respiratory diseases including pneumonia, foot rot, internal parasites, pregnancy toxosis and feed toxicity. Feed toxicity can vary based on breed and location. Certain foreign fruits and vegetables can be toxic to different breeds of goats.

Goats can become infected with various viral and bacterial diseases, such as foot-and-mouth disease, caprine arthritis encephalitis, caseous lymphadenitis, pinkeye, mastitis, and pseudorabies. They can transmit a number of zoonotic diseases to people, such as tuberculosis, brucellosis, Q-fever, and rabies.

## Life Expectancy

Life expectancy for goats is between 15 and 18 years. An instance of a goat reaching the age of 24 has been reported.

Several factors can reduce this average expectancy; problems during kidding can lower a doe's expected life span to 10 or 11, and stresses of going into rut can lower a buck's expected life span to eight to 10 years.

## Agriculture

Goat husbandry is common through the Norte Chico region in Chile. Intensive goat husbandry in drylands may produce severe erosion and desertification. Image from upper Limarí River.

A goat is useful to humans when it is living and when it is dead, first as a renewable provider of milk, manure, and fiber, and then as meat and hide. Some charities provide goats to impoverished people in poor countries, because goats are easier and cheaper to manage than cattle, and have multiple uses. In addition, goats are used for driving and packing purposes.

The Boer goat – in this case a buck – is a widely kept meat breed.

The intestine of goats is used to make "catgut", which is still in use as a material for internal human surgical sutures and strings for musical instruments. The horn of the goat, which signifies plenty and wellbeing (the cornucopia), is also used to make spoons.

## Worldwide Goat Population Statistics

According to the Food and Agriculture Organization (FAO), the top producers of goat milk in 2008 were India (4 million metric tons), Bangladesh (2.16 million metric tons) and the Sudan (1.47 million metric tons). India slaughters 41% of 124.4 million goats each year. The 0.6 million metric tones of goat meat make up 8% of India's annual meat production.

## Husbandry

Husbandry, or animal care and use, varies by region and culture. The particular housing used for goats depends not only on the intended use of the goat, but also on the region of the world where they are raised. Historically, domestic goats were generally kept in herds that wandered on hills or other grazing areas, often tended by goatherds who were frequently children or adolescents, similar to the more widely known shepherd. These methods of herding are still used today.

Reared goat (Husbandry).

In some parts of the world, especially Europe and North America, distinct breeds of goats are kept for dairy (milk) and for meat production. Excess male kids of dairy breeds are typically slaughtered for meat. Both does and bucks of meat breeds may be slaughtered for meat, as well as older animals of any breed. The meat of older bucks (more than one year old) is generally considered not desirable for meat for human consumption. Castration at a young age prevents the development of typical buck odor.

For smallholder farmers in many countries, such as this woman
from Burkina Faso, goats are important livestock.

Dairy goats are generally pastured in summer and may be stabled during the winter. As dairy does are milked daily, they are generally kept close to the milking shed. Their grazing is typically supplemented with hay and concentrates. Stabled goats may be kept in stalls similar to horses, or in larger group pens. In the US system, does are generally rebred annually. In some European commercial dairy systems, the does are bred only twice, and are milked continuously for several years after the second kidding.

Meat goats are more frequently pastured year-round, and may be kept many miles from barns. Angora and other fiber breeds are also kept on pasture or range. Range-kept and pastured goats may be supplemented with hay or concentrates, most frequently during the winter or dry seasons.

In the Indian subcontinent and much of Asia, goats are kept largely for milk production, both in commercial and household settings. The goats in this area may be kept closely housed or may be allowed to range for fodder. The Salem Black goat is herded to pasture in fields and along roads during the day, but is kept penned at night for safe-keeping.

In Africa and the Mideast, goats are typically run in flocks with sheep. This maximizes the production per acre, as goats and sheep prefer different food plants. Multiple types of goat-raising are found in Ethiopia, where four main types have been identified: pastured in annual crop systems, in perennial crop systems, with cattle, and in arid areas, under pastoral (nomadic) herding systems. In all four systems, however, goats were typically kept in extensive systems, with few purchased inputs. Household goats are traditionally kept in Nigeria. While many goats are allowed to wander the homestead or village, others are kept penned and fed in what is called a 'cut-and-carry' system. This type of husbandry is also used in parts of Latin America. Cut-and-carry, which refers to the practice of cutting down grasses, corn or cane for feed rather than allowing the animal access to the field, is particularly suited for types of feed, such as corn or cane, that are easily destroyed by trampling.

Pet goats may be found in many parts of the world when a family keeps one or more animals for emotional reasons rather than as production animals. It is becoming more common for goats to be kept exclusively as pets in North America and Europe.

## Meat

The taste of goat kid meat is similar to that of spring lamb meat; in fact, in the English-speaking islands of the Caribbean, and in some parts of Asia, particularly Bangladesh, Pakistan and India, the word "mutton" is used to describe both goat and sheep meat. However, some compare the taste of goat meat to veal or venison, depending on the age and condition of the goat. Its flavor is said to be primarily linked to the presence of 4-methyloctanoic and 4-methylnonanoic acid. It can be prepared in a variety of ways, including stewing, baking, grilling, barbecuing, canning, and frying; it can be minced, curried, or made into sausage. Due to its low fat content, the meat can toughen at high temperatures if cooked without additional moisture. One of the most popular goats grown for meat is the South African Boer, introduced into the United States in the early 1990s. The New Zealand Kiko is also considered a meat breed, as is the myotonic or "fainting goat", a breed originating in Tennessee.

## Milk, Butter and Cheese

Goats produce about 2% of the world's total annual milk supply. Some goats are bred specifically for milk. If the strong-smelling buck is not separated from the does, his scent will affect the milk.

Goat milk naturally has small, well-emulsified fat globules, which means the cream remains suspended in the milk, instead of rising to the top, as in raw cow milk; therefore, it does not need to be homogenized. Indeed, if the milk is to be used to make cheese, homogenization is not recommended, as this changes the structure of the milk, affecting the culture's ability to coagulate the milk and the final quality and yield of cheese.

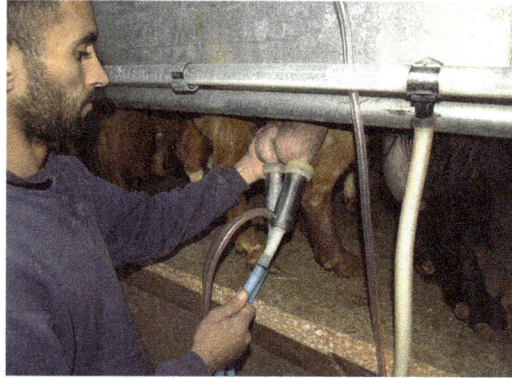

A goat being machine milked on an organic farm.

Dairy goats in their prime (generally around the third or fourth lactation cycle) average—2.7 to 3.6 kg (6 to 8 lb)—of milk production daily—roughly 2.8 to 3.8 l (3 to 4 U.S. qt)—during a ten-month lactation, producing more just after freshening and gradually dropping in production toward the end of their lactation. The milk generally averages 3.5% butterfat.

Goat milk is commonly processed into cheese, butter, ice cream, yogurt, *cajeta* and other products. Goat cheese is known as *fromage de chèvre* ("goat cheese") in France. Some varieties include Rocamadour and Montrachet. Goat butter is white because goats produce milk with the yellow beta-carotene converted to a colorless form of vitamin A.

## Nutrition

The American Academy of Pediatrics discourages feeding infants milk derived from goats. An April 2010 case report summarizes their recommendation and presents "a comprehensive review of the consequences associated with this dangerous practice", also stating, "Many infants are exclusively fed unmodified goat's milk as a result of cultural beliefs as well as exposure to false online information. Anecdotal reports have described a host of morbidities associated with that practice, including severe electrolyte abnormalities, metabolic acidosis, megaloblastic anemia, allergic reactions including life-threatening anaphylactic shock, hemolytic uremic syndrome, and infections." Untreated caprine brucellosis results in a 2% case fatality rate. According to the USDA, doe milk is not recommended for human infants because it contains "inadequate quantities of iron, folate, vitamins C and D, thiamine, niacin, vitamin $B_6$, and pantothenic acid to meet an infant's nutritional needs" and may cause harm to an infant's kidneys and could cause metabolic damage.

The department of health in the United Kingdom has repeatedly released statements stating on various occasions that "Goats' milk is not suitable for babies, and infant formulas and follow-on formulas based on goats' milk protein have not been approved for use in Europe", and "infant milks based on goats' milk protein are not suitable as a source of nutrition for infants." Moreover, according to the Canadian federal health department *Health Canada*, most of the dangers of, and counter-indications for, feeding unmodified goat's milk to infants parallel those associated with unmodified cow's milk—especially insofar as allergic reactions go.

However, some farming groups promote the practice. For example, Small Farm Today, in 2005, claimed beneficial use in invalid and convalescent diets, proposing that glycerol ethers, possibly important in nutrition for nursing infants, are much higher in does' milk than in cows' milk. A 1970 book on animal breeding claimed that does' milk differs from cows' or humans' milk by having higher digestibility, distinct alkalinity, higher buffering capacity, and certain therapeutic values in human medicine and nutrition. George Mateljan suggested doe milk can replace ewe milk or cow milk in diets of those who are allergic to certain mammals' milk. However, like cow milk, doe milk has lactose (sugar), and may cause gastrointestinal problems for individuals with lactose intolerance. In fact, the level of lactose is similar to that of cow milk.

Some researchers and companies producing goat's milk products have made claims that goat's milk is better for human health than most Western cow's milk due to it mostly lacking a form of β-casein proteins called A1, and instead mostly containing the A2 form, which does not metabolize to β-casomorphin 7 in the body.

Table: Basic Composition of Various Milks (Mean Values Per 100 G).

| Constituent | Doe (goat) | Cow | Human |
|---|---|---|---|
| Fat (g) | 3.8 | 3.6 | 4.0 |
| Protein (g) | 3.5 | 3.3 | 1.2 |
| Lactose (g) | 4.1 | 4.6 | 6.9 |
| Ash (g) | 0.8 | 0.7 | 0.2 |
| Total solids (g) | 12.2 | 12.3 | 12.3 |
| Calories | 70 | 69 | 68 |

Table: Milk Composition Analysis, Per 100 Grams.

| Constituents | Unit | Cow | Doe (goat) | Ewe (sheep) | Water buffalo |
|---|---|---|---|---|---|
| Water | g | 87.8 | 88.9 | 83.0 | 81.1 |
| Protein | g | 3.2 | 3.1 | 5.4 | 4.5 |
| Fat | g | 3.9 | 3.5 | 6.0 | 8.0 |
| Carbohydrates | g | 4.8 | 4.4 | 5.1 | 4.9 |
| Energy | kcal | 66 | 60 | 95 | 110 |
| Energy | kJ | 275 | 253 | 396 | 463 |
| Sugars (lactose) | g | 4.8 | 4.4 | 5.1 | 4.9 |
| Cholesterol | mg | 14 | 10 | 11 | 8 |
| Calcium | IU | 120 | 100 | 170 | 195 |
| Saturated fatty acids | g | 2.4 | 2.3 | 3.8 | 4.2 |
| Monounsaturated fatty acids | g | 1.1 | 0.8 | 1.5 | 1.7 |
| Polyunsaturated fatty acids | g | 0.1 | 0.1 | 0.3 | 0.2 |

These compositions vary by breed (especially in the Nigerian Dwarf breed), animal, and point in the lactation period.

## Fiber

An Angora goat.

The Angora breed of goats produces long, curling, lustrous locks of mohair. The entire body of the goat is covered with mohair and there are no guard hairs. The locks constantly grow to four inches or more in length. Angora crossbreeds, such as the pygora and the nigora, have been created to produce mohair and/or cashgora on a smaller, easier-to-manage animal. The wool is shorn twice a year, with an average yield of about 4.5 kg (10 lb).

Most goats have softer insulating hairs nearer the skin, and longer guard hairs on the surface. The desirable fiber for the textile industry is the former, and it goes by several names (down, cashmere and pashmina). The coarse guard hairs are of little value as they are too coarse, difficult to spin and difficult to dye. The cashmere goat produces a commercial quantity of cashmere wool, which is one of the most expensive natural fibers commercially produced; cashmere is very fine and soft. The cashmere goat fiber is harvested once a year, yielding around 260 g (9 oz) of down.

In South Asia, cashmere is called "pashmina" (from Persian *pashmina*, "fine wool"). In the 18th and early 19th centuries, Kashmir (then called Cashmere by the British), had a thriving industry producing shawls from goat-hair imported from Tibet and Tartary through Ladakh. The shawls were introduced into Western Europe when the General in Chief of the French campaign in Egypt sent one to Paris. Since these shawls were produced in the upper Kashmir and Ladakh region, the wool came to be known as "cashmere".

## Land Clearing

Goats have been used by humans to clear unwanted vegetation for centuries. They have been described as "eating machines" and "biological control agents". There has been a resurgence of this in North America since 1990, when herds were used to clear dry brush from California hillsides thought to be endangered by potential wildfires. This form of using goats to clear land is sometimes known as conservation grazing. Since then, numerous public and private agencies have hired private herds from companies such as Rent A Goat to perform similar tasks. This may be expensive and their smell may be a nuisance. This practice has become popular in the Pacific Northwest, where they are used to remove invasive species not easily removed by humans, including (thorned) blackberry vines and poison oak. Chattanooga, TN and Spartanburg, SC have used goats to control kudzu, an invasive plant species prevalent in the southeastern United States.

Goats managing the landscape alongside German autobahn A59.

## Use for Medical Training

As a goat's anatomy and physiology is not too dissimilar from that of humans, some countries' militaries use goats to train combat medics. In the United States, goats have become the main animal species used for this purpose after the Pentagon phased out using dogs for medical training in the 1980s. While modern mannequins used in medical training are quite efficient in simulating the behavior of a human body, trainees feel that "the goat exercise provides a sense of urgency that only real life trauma can provide".

## As Pets

Some people choose goats as a pet because of their ability to form close bonds with their human guardians. Because of goats' herd mentality, they will follow their owners around and form close bonds with them.

## Breeds

Goat breeds fall into overlapping, general categories. They are generally distributed in those used for dairy, fiber, meat, skins, and as companion animals. Some breeds are also particularly noted as pack goats.

## Showing

Goat breeders' clubs frequently hold shows, where goats are judged on traits relating to conformation, udder quality, evidence of high production, longevity, build and muscling (meat goats and pet goats) and fiber production and the fiber itself (fiber goats). People who show their goats usually keep registered stock and the offspring of award-winning animals command a higher price. Registered goats, in general, are usually higher-priced if for no other reason than that records have been kept proving their ancestry and the production and other data of their sires, dams, and other ancestors. A registered doe is usually less of a gamble than buying a doe at random (as at an auction or sale barn) because of these records and the reputation of the breeder. Children's clubs such as 4-H also allow goats to be shown. Children's shows often include a showmanship class, where the cleanliness and presentation of both the animal and the exhibitor as well as the handler's ability and skill in handling the goat are scored. In a showmanship class, conformation is irrelevant since this is not what is being judged.

A Nigerian Dwarf milker in show clip. This doe is angular and dairy with a capacious and well supported mammary system.

Various "Dairy Goat Scorecards" (milking does) are systems used for judging shows in the US. The American Dairy Goat Association (ADGA) scorecard for an adult doe includes a point system of a hundred total with major categories that include general appearance, the dairy character of a doe (physical traits that aid and increase milk production), body capacity, and specifically for the mammary system. Young stock and bucks are judged by different scorecards which place more emphasis on the other three categories; general appearance, body capacity, and dairy character.

The American Goat Society (AGS) has a similar, but not identical scorecard that is used in their shows. The miniature dairy goats may be judged by either of the two scorecards. The "Angora Goat scorecard" used by the Colored Angora Goat Breeder's Association (CAGBA), which covers the white and the colored goats, includes evaluation of an animal's fleece color, density, uniformity, fineness, and general body confirmation. Disqualifications include: a deformed mouth, broken down pasterns, deformed feet, crooked legs, abnormalities of testicles, missing testicles, more than 3 inch split in scrotum, and close-set or distorted horns.

## Feral Goats

Feral goat in Aruba.

Goats readily revert to the wild (become feral) if given the opportunity. The only domestic animal known to return to feral life as swiftly is the cat. Feral goats have established themselves in many areas: they occur in Australia, New Zealand, Great Britain, the Galapagos and in many other places. When feral goats reach large populations in habitats which provide unlimited water supply and which do not contain sufficient large predators or which are otherwise vulnerable to goats' aggressive grazing habits, they may have serious effects, such as removing native scrub, trees and other vegetation which is required by a wide range of other creatures, not just other grazing or browsing animals. Feral goats are common in Australia. However, in other circumstances where predator pressure is maintained, they may be accommodated into some balance in the local food web.

# Donkeys

The donkey or ass (Equus africanus asinus) is a domesticated member of the horse family, Equidae. The wild ancestor of the donkey is the African wild ass, E. africanus. The donkey has been used as a working animal for at least 5000 years. There are more than 40 million donkeys in the world, mostly in underdeveloped countries, where they are used principally as draught or pack animals. Working donkeys are often associated with those living at or below subsistence levels. Small numbers of donkeys are kept for breeding or as pets in developed countries.

A male donkey or ass is called a jack, a female a jenny or jennet; a young donkey is a foal. Jack donkeys are often used to mate with female horses to produce mules; the biological "reciprocal" of a mule, from a stallion and jenny as its parents instead, is called a hinny.

Asses were first domesticated around 3000 BC, probably in Egypt or Mesopotamia, and have spread around the world. They continue to fill important roles in many places today. While domesticated species are increasing in numbers, the African wild ass is an endangered species. As beasts of burden and companions, asses and donkeys have worked together with humans for millennia.

## Scientific and Common Names

Traditionally, the scientific name for the donkey is Equus asinus asinus based on the principle of priority used for scientific names of animals. However, the International Commission on Zoological Nomenclature ruled in 2003 that if the domestic species and the wild species are considered subspecies of one another, the scientific name of the wild species has priority, even when that subspecies was described after the domestic subspecies. This means that the proper scientific name for the donkey is Equus africanus asinus when it is considered a subspecies, and Equus asinus when it is considered a species.

At one time, the synonym *ass* was the more common term for the donkey. The first recorded use of *donkey* was in either 1784 or 1785. While the word *ass* has cognates in most other Indo-European languages, *donkey* is an etymologically obscure word for which no credible cognate has been identified. Hypotheses on its derivation include the following:

- Perhaps from Spanish, for its don-like gravity; the donkey was also known as "the King of Spain's trumpeter".

- Perhaps a diminutive of *dun* (dull grayish-brown), a typical donkey colour.

- Perhaps from the name *Duncan*.

- Perhaps of imitative origin.

From the 18th century, *donkey* gradually replaced *ass*, and *jenny* replaced *she-ass*, which is now considered archaic. The change may have come about through a tendency to avoid pejorative terms in speech, and be comparable to the substitution in North American English of *rooster* for *cock*, or that of *rabbit* for *coney*, which was formerly homophonic with *cunny*. By the end of the 17th century, changes in pronunciation of both *ass* and *arse* had caused them to become homophones. Other words used for the ass in English from this time include *cuddy* in Scotland, *neddy* in southwest England and *dicky* in the southeast; *moke* is documented in the 19th century, and may be of Welsh or Gypsy origin.

## Characteristics

Classic British seaside donkeys in Skegness.

Donkeys vary considerably in size, depending on breed and management. The height at the withers ranges from 7.3 to 15.3 hands (31 to 63 inches, 79 to 160 cm), and the weight from 80 to 480 kg (180 to 1,060 lb). Working donkeys in the poorest countries have a life expectancy of 12 to 15 years; in more prosperous countries, they may have a lifespan of 30 to 50 years.

Donkeys are adapted to marginal desert lands. Unlike wild and feral horses, wild donkeys in dry areas are solitary and do not form harems. Each adult donkey establishes a home range; breeding over a large area may be dominated by one jack. The loud call or bray of the donkey, which typically lasts for twenty seconds and can be heard for over three kilometres, may help keep in contact with other donkeys over the wide spaces of the desert. Donkeys have large ears, which may pick up more distant sounds, and may help cool the donkey's blood. Donkeys can defend themselves by biting, striking with the front hooves or kicking with the hind legs.

## Breeding

A jenny is normally pregnant for about 12 months, though the gestation period varies from 11 to 14 months, and usually gives birth to a single foal. Births of twins are rare, though less so than in horses. About 1.7 percent of donkey pregnancies result in twins; both foals survive in about 14 percent of those. In general jennies have a conception rate that is lower than that of horses (*i.e.* less than the 60–65% rate for mares).

A 3-week-old donkey.

Although jennies come into heat within 9 or 10 days of giving birth, their fertility remains low, and it is likely the reproductive tract has not returned to normal. Thus it is usual to wait one or two further oestrous cycles before rebreeding, unlike the practice with mares. Jennies are usually very protective of their foals, and some will not come into estrus while they have a foal at side. The time lapse involved in rebreeding, and the length of a jenny's gestation, means that a jenny will have fewer than one foal per year. Because of this and the longer gestation period, donkey breeders do not expect to obtain a foal every year, as horse breeders often do, but may plan for three foals in four years.

Donkeys can interbreed with other members of the family Equidae, and are commonly interbred with horses. The hybrid between a jack and a mare is a mule, valued as a working and riding animal in many countries. Some large donkey breeds such as the Asino di Martina Franca, the Baudet de Poitou and the Mammoth Jack are raised only for mule production. The hybrid between a stallion and a jenny is a hinny, and is less common. Like other inter-species hybrids, mules and hinnies are usually sterile. Donkeys can also breed with zebras in which the offspring is called a zonkey (among other names).

## Behavior

Donkeys have a notorious reputation for stubbornness, but this has been attributed to a much stronger sense of self-preservation than exhibited by horses. Likely based on a stronger prey instinct and a weaker connection with humans, it is considerably more difficult to force or frighten a donkey into doing something it perceives to be dangerous for whatever reason. Once a person has earned their confidence they can be willing and companionable partners and very dependable in work.

Although formal studies of their behavior and cognition are rather limited, donkeys appear to be quite intelligent, cautious, friendly, playful, and eager to learn.

## Uses

### Economic Use

The donkey has been used as a working animal for at least 5000 years. Of the more than 40 million donkeys in the world, about 96% are in underdeveloped countries, where they are used principally as pack animals or for draught work in transport or agriculture. After human labour, the donkey is the cheapest form of agricultural power. They may also be ridden, or used for threshing, raising water, milling and other work. Working donkeys are often associated with those living at or below subsistence levels. Some cultures that prohibit women from working with oxen in agriculture do not extend this taboo to donkeys, allowing them to be used by both sexes.

Donkeys bring supplies through the jungle to a camp outpost in Tayrona National Natural Park in northern Colombia.

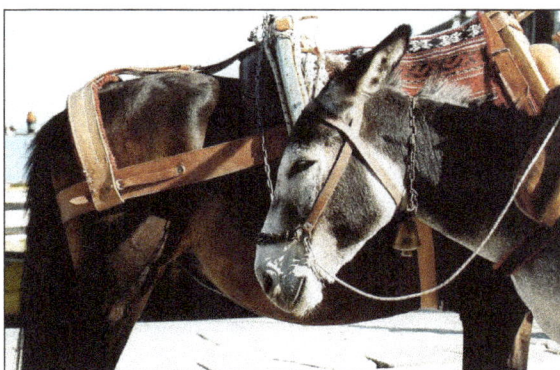

On the island of Hydra, because cars are outlawed, donkeys and mules are virtually the only ways to transport heavy goods.

In developed countries where their use as beasts of burden has disappeared, donkeys are used to sire mules, to guard sheep, for donkey rides for children or tourists, and as pets. Donkeys may be pastured or stabled with horses and ponies, and are thought to have a calming effect on nervous horses. If a donkey is introduced to a mare and foal, the foal may turn to the donkey for support after it has been weaned from its mother.

A few donkeys are milked or raised for meat; in Italy, which has the highest consumption of equine meat in Europe and where donkey meat is the main ingredient of several regional dishes, about 1000 donkeys were slaughtered in 2010, yielding approximately 100 tonnes of meat. Asses' milk may command good prices: the average price in Italy in 2009 was €15 per litre, and a price of €6 per 100 ml was reported from Croatia in 2008; it is used for soaps and cosmetics as well as dietary purposes. The niche markets for both milk and meat are expanding. In the past, donkey skin was used in the production of parchment. In 2017, the UK based charity The Donkey Sanctuary estimated that 1.8 million skins were traded every year, but the demand could be as high as 10 million.

In China, donkey meat is considered a delicacy with some restaurants specializing in such dishes, and Guo Li Zhuang restaurants offer the genitals of donkeys in dishes. Donkey-hide gelatin is produced by soaking and stewing the hide to make a traditional Chinese medicine product. Ejiao, the gelatine produced by boiling donkey skins, can sell for up to $388 per kilo, at October 2017 prices.

In 2017, a drop in the number of Chinese donkeys, combined with the fact that they are slow to reproduce, meant that Chinese suppliers began to look to Africa. As a result of the increase in demand, and the price that could be charged, Kenya opened three donkey abattoirs. Concerns for donkeys' well-being, however, have resulted in a number of African countries (including Uganda, Tanzania, Botswana, Niger, Burkina Faso, Mali, and Senegal) banning China from buying their donkey products.

## Care

## Shoeing

Donkey hooves are more elastic than those of horses, and do not naturally wear down as fast. Regular clipping may be required; neglect can lead to permanent damage. Working donkeys may need to be shod. Donkey shoes are similar to horseshoes, but usually smaller and without toe-clips.

A donkey shoe with calkins.

Farriers shoeing a donkey in Cyprus in 1900.

## Nutrition

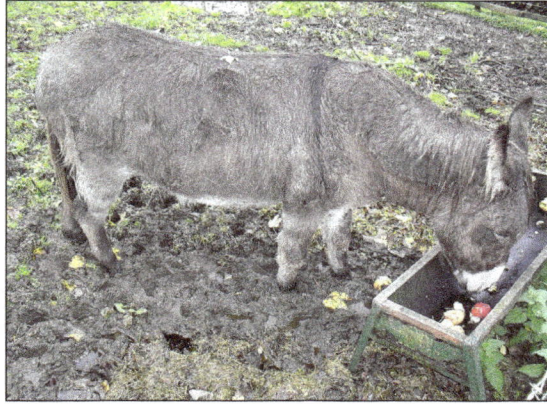

Donkey eating apples from a trough.

In their native arid and semi-arid climates, donkeys spend more than half of each day foraging and feeding, often on poor quality scrub. The donkey has a tough digestive system in which roughage is efficiently broken down by hind gut fermentation, microbial action in the caecum and large intestine. While there is no marked structural difference between the gastro-intestinal tract of a donkey and that of a horse, the digestion of the donkey is more efficient. It needs less food than a horse or pony of comparable height and weight, approximately 1.5 percent of body weight per day in dry matter, compared to the 2–2.5 percent consumption rate possible for a horse. Donkeys are also less prone to colic. The reasons for this difference are not fully understood; the donkey may have different intestinal flora to the horse, or a longer gut retention time.

Poitou donkeys.

Donkeys obtain most of their energy from structural carbohydrates. Some suggest that a donkey needs to be fed only straw (preferably barley straw), supplemented with controlled grazing in the summer or hay in the winter, to get all the energy, protein, fat and vitamins it requires; others recommend some grain to be fed, particularly to working animals, and others advise against feeding straw. They do best when allowed to consume small amounts of food over long periods. They can meet their nutritional needs on 6 to 7 hours of grazing per day on average dryland pasture that is not stressed by drought. If they are worked long hours or do not have access to pasture, they require hay or a similar dried forage, with no more than a 1:4 ratio of legumes to grass. They also require salt and mineral supplements, and access to clean, fresh water. In temperate climates the forage available is often too abundant and too rich; over-feeding may cause weight gain and obesity, and lead to metabolic disorders such as founder (laminitis) and hyperlipaemia, or to gastric ulcers.

Throughout the world, working donkeys are associated with the very poor, with those living at or below subsistence level. Few receive adequate food, and in general donkeys throughout the Third World are under-nourished and over-worked.

## Burro

A burro pulling a cart during the Carnival of Huejotzingo.

In the Iberian Peninsula and the Americas, a *burro* is a small donkey. The Domestic Animal Diversity Information System (DAD-IS) of the FAO lists the burro as a specific breed of ass. In Mexico, the donkey population is estimated at three million. There are also substantial *burro* populations in El Salvador, Guatemala, and Nicaragua.

Burro is the Spanish and Portuguese word for donkey. In Spanish, burros may also be called burro mexicano ('Mexican donkey'), burro criollo ('Criollo donkey'), or burro criollo mexicano. In the United States, "burro" is used as a loan word by English speakers to describe any small donkey used primarily as a pack animal, as well as to describe the feral donkeys that live in Arizona, California, Oregon, Utah, Texas and Nevada.

Among donkeys, burros tend to be on the small side. A study of working burros in central Mexico found a weight range of 50–186 kilograms (110–410 lb), with an average weight of 122 kg (269 lb) for males and 112 kg (247 lb) for females. Height at the withers varied from 87–120 cm (34–47 in), with an average of approximately 108 cm (43 in), and girth measurements ranged from 88–152 cm (35–60 in), with an average of about 120 cm (47 in). The average age of the burros in the study was 6.4 years; evaluated by their teeth, they ranged from 1 to 17 years old. They are gray in color. Mexican burros tend to be smaller than their counterparts in the USA, which are both larger and more robust. To strengthen their bloodstock, in May 2005, the state of Jalisco imported 11 male and female donkeys from Kentucky.

## Donkey Hybrids

A male donkey (jack) can be crossed with a female horse to produce a mule. A male horse can be crossed with a female donkey (jenny) to produce a hinny.

Horse-donkey hybrids are almost always sterile because horses have 64 chromosomes whereas donkeys have 62, producing offspring with 63 chromosomes. Mules are much more common than hinnies. This is believed to be caused by two factors, the first being proven in cat hybrids, that when the chromosome count of the male is the higher, fertility rates drop. The lower progesterone production of the jenny may also lead to early embryonic loss. In addition, there are reasons not directly related to reproductive biology. Due to different mating behavior, jacks are often more willing to cover mares than stallions are to breed jennies. Further, mares are usually larger than jennies and thus have more room for the ensuing foal to grow in the womb, resulting in a larger animal at birth. It is commonly believed that mules are more easily handled and also physically stronger than hinnies, making them more desirable for breeders to produce.

The offspring of a zebra-donkey cross is called a zonkey, zebroid, zebrass, or zedonk; zebra mule is an older term, but still used in some regions today. The foregoing terms generally refer to hybrids produced by breeding a male zebra to a female donkey. Zebra hinny, zebret and zebrinny all refer to the cross of a female zebra with a male donkey. Zebrinnies are rarer than zedonkies because female zebras in captivity are most valuable when used to produce full-blooded zebras. There are not enough female zebras breeding in captivity to spare them for hybridizing; there is no such limitation on the number of female donkeys breeding.

# Camels

A camel is an even-toed ungulate in the genus *Camelus* that bears distinctive fatty deposits known as "humps" on its back. Camels have long been domesticated and, as livestock, they provide food (milk and meat) and textiles (fiber and felt from hair). As working animals, camels—which are uniquely suited to their desert habitats—are a vital means of transport for passengers and cargo. There are three surviving species of camel. The one-humped dromedary makes up 94% of the world's camel population, and the two-humped Bactrian camel makes up the remainder. The Wild Bactrian camel is a separate species and is now critically endangered.

The dromedary (C. dromedarius), also known as the Arabian camel, inhabits the Middle East and the Horn of Africa, while the Bactrian (C. bactrianus) inhabits Central Asia, including the historical region of Bactria. The critically endangered wild Bactrian (C. ferus) is found only in remote areas of northwest China and Mongolia. An extinct species of camel in the separate genus Camelops, known as C. hesternus, lived in western North America until humans entered the continent at the end of the Pleistocene.

## Biology

The average life expectancy of a camel is 40 to 50 years. A full-grown adult camel stands 1.85 m (6 ft 1 in) at the shoulder and 2.15 m (7 ft 1 in) at the hump. Camels can run at up to 65 km/h (40 mph) in short bursts and sustain speeds of up to 40 km/h (25 mph). Bactrian camels weigh 300 to 1,000 kg (660 to 2,200 lb) and dromedaries 300 to 600 kg (660 to 1,320 lb). The widening toes on a camel's hoof provide supplemental grip for varying soil sediments.

The male dromedary camel has an organ called a dulla in its throat, a large, inflatable sac he extrudes from his mouth when in rut to assert dominance and attract females. It resembles a long, swollen, pink tongue hanging out of the side of its mouth. Camels mate by having both male and female sitting on the ground, with the male mounting from behind. The male usually ejaculates three or four times within a single mating session. Camelids are the only ungulates to mate in a sitting position.

## Ecological and Behavioral Adaptations

Camels do not directly store water in their humps; they are reservoirs of fatty tissue. Concentrating body fat in their humps minimizes the insulating effect fat would have if distributed over the rest of their bodies, helping camels survive in hot climates. When this tissue is metabolized, it yields more than one gram of water for every gram of fat processed. This fat metabolization, while releasing energy, causes water to evaporate from the lungs during respiration (as oxygen is required for the metabolic process): overall, there is a net decrease in water.

A camel's thick coat is one of its many adaptations that aid it in desert-like conditions.

Somalia has the world's largest population of camels.

Camels have a series of physiological adaptations that allow them to withstand long periods of time without any external source of water. The dromedary camel can drink as seldom as once every 10 days even under very hot conditions, and can lose up to 30% of its body mass due to dehydration. Unlike other mammals, camels' red blood cells are oval rather than circular in shape. This facilitates the flow of red blood cells during dehydration and makes them better at withstanding high osmotic variation without rupturing when drinking large amounts of water: a 600 kg (1,300 lb) camel can drink 200 L (53 US gal) of water in three minutes.

Camels are able to withstand changes in body temperature and water consumption that would kill most other animals. Their temperature ranges from 34 °C (93 °F) at dawn and steadily increases to 40 °C (104 °F) by sunset, before they cool off at night again. In general, to compare between camels and the other livestock, camels lose only 1.3 liters of fluid intake every day while the other livestock lose 20 to 40 liters per day. Maintaining the brain temperature within certain limits is critical for animals; to assist this, camels have a rete mirabile, a complex of arteries and veins lying very close to each other which utilizes countercurrent blood flow to cool blood flowing to the brain. Camels rarely sweat, even when ambient temperatures reach 49 °C (120 °F). Any sweat that does occur evaporates at the skin level rather than at the surface of their coat; the heat of vaporization therefore comes from body heat rather than ambient heat. Camels can withstand losing 25% of their body weight to sweating, whereas most other mammals can withstand only about 12–14% dehydration before cardiac failure results from circulatory disturbance.

When the camel exhales, water vapor becomes trapped in their nostrils and is reabsorbed into the body as a means to conserve water. Camels eating green herbage can ingest sufficient moisture in milder conditions to maintain their bodies' hydrated state without the need for drinking.

Domesticated camel calves lying in sternal recumbency, a position that aids heat loss.

The camels' thick coats insulate them from the intense heat radiated from desert sand; a shorn camel must sweat 50% more to avoid overheating. During the summer the coat becomes lighter in color, reflecting light as well as helping avoid sunburn. The camel's long legs help by keeping its body farther from the ground, which can heat up to 70 °C (158 °F). Dromedaries have a pad of thick tissue over the sternum called the *pedestal*. When the animal lies down in a sternal recumbent position, the pedestal raises the body from the hot surface and allows cooling air to pass under the body.

Camels' mouths have a thick leathery lining, allowing them to chew thorny desert plants. Long eyelashes and ear hairs, together with nostrils that can close, form a barrier against sand. If sand gets lodged in their eyes, they can dislodge it using their transparent third eyelid. The camels' gait and widened feet help them move without sinking into the sand.

The kidneys and intestines of a camel are very efficient at reabsorbing water. Camels' kidneys have a 1:4 cortex to medulla ratio. Thus, the medullary part of a camel's kidney occupies twice as much area as a cow's kidney. Secondly, renal corpuscles have a smaller diameter, which reduces surface area for filtration. These two major anatomical characteristics enable camels to conserve water and limit the volume of urine in extreme desert conditions. Camel urine comes out as a thick syrup, and camel faeces are so dry that they do not require drying when the Bedouins use them to fuel fires. The camels are able to live in difficult conditions without drinking water due to their ability to produce small and dry droppings as well as they use the water to maintain their body's temperature to fit with the region surrounding them.

The camel immune system differs from those of other mammals. Normally, the Y-shaped antibody molecules consist of two heavy (or long) chains along the length of the Y, and two light (or short) chains at each tip of the Y. Camels, in addition to these, also have antibodies made of only two heavy chains, a trait that makes them smaller and more durable. These "heavy-chain-only" antibodies, discovered in 1993, are thought to have developed 50 million years ago, after camelids split from ruminants and pigs.

## Genetics

Caravan of dromedaries, Giza pyramid complex, Egypt.

The karyotypes of different camelid species have been studied earlier by many groups, but no agreement on chromosome nomenclature of camelids has been reached. A 2007 study flow sorted camel chromosomes, building on the fact that camels have 37 pairs of chromosomes (2n=74), and found that the karyotype consisted of one metacentric, three submetacentric, and 32 acrocentric autosomes. The Y is a small metacentric chromosome, while the X is a large metacentric chromosome.

Skull of an F1 hybrid camel, Museum of Osteology, Oklahoma.

The hybrid camel, a hybrid between Bactrian and dromedary camels, has one hump, though it has an indentation 4–12 cm (1.6–4.7 in) deep that divides the front from the back. The hybrid is 2.15

m (7 ft 1 in) at the shoulder and 2.32 m (7 ft 7 in) tall at the hump. It weighs an average of 650 kg (1,430 lb) and can carry around 400 to 450 kg (880 to 990 lb), which is more than either the dromedary or Bactrian can.

According to molecular data, the New World and Old World camelids diverged about 11 million years ago. In spite of this, these species can hybridize and produce viable offspring. The cama is a camel-llama hybrid bred by scientists to see how closely related the parent species are. Scientists collected semen from a camel via an artificial vagina and inseminated a llama after stimulating ovulation with gonadotrophin injections. The cama is halfway in size between a camel and a llama and lacks a hump. It has ears intermediate between those of camels and llamas, longer legs than the llama, and partially cloven hooves. Like the mule, camas are sterile, despite both parents having the same number of chromosomes. The wild Bactrian camel (*C. ferus*) separated from the domestic Bactrian camel (*C. bactrianus*) about 1 million years ago.

## Domestication

Like horses before their extinction in their continent of origin, camels spread across the Bering land bridge, moving in the opposite direction from the Asian immigration to America. They survived in the Old World, and eventually humans domesticated them and spread them globally. Along with many other megafauna in North America, the original wild camels were wiped out during the spread of Native Americans from Asia into North America, 12,000 to 10,000 years ago. Most camels surviving today are domesticated. Although feral populations exist in Australia, India and Kazakhstan, wild camels survive only in the wild Bactrian camel population of the Gobi Desert.

Humans may have first domesticated dromedaries in Somalia and southern Arabia around 3,000 BC, and Bactrian camels in central Asia around 2,500 BC, as at Shahr-e Sukhteh (also known as the Burnt City), Iran.

Martin Heide's 2010 work on the domestication of the camel tentatively concludes that humans had domesticated the Bactrian camel by at least the middle of the third millennium somewhere east of the Zagros Mountains, with the practice then moving into Mesopotamia. Heide suggests that mentions of camels "in the patriarchal narratives may refer, at least in some places, to the Bactrian camel", while noting that the camel is not mentioned in relationship to Canaan.

Recent excavations in the Timna Valley by Lidar Sapir-Hen and Erez Ben-Yosef discovered what may be the earliest domestic camel bones yet found in Israel or even outside the Arabian Peninsula, dating to around 930 BC. This garnered considerable media coverage, as it was described as evidence that the stories of Abraham, Jacob, Esau, and Joseph were written after this time.

The existence of camels in Mesopotamia—but not in the eastern Mediterranean lands—is not a new idea.

The official report by Sapir-Hen and Ben-Joseph notes:

> "The introduction of the dromedary camel (Camelus dromedarius) as a pack animal to the southern Levant substantially facilitated trade across the vast deserts of Arabia, promoting both economic and social change. This has generated extensive discussion regarding the date of the earliest domestic camel in the southern Levant (and beyond). Most scholars

today agree that the dromedary was exploited as a pack animal sometime in the early Iron Age and concludes."

Current data from copper smelting sites of the Aravah Valley enable us to pinpoint the introduction of domestic camels to the southern Levant more precisely based on stratigraphic contexts associated with an extensive suite of radiocarbon dates. The data indicate that this event occurred not earlier than the last third of the 10th century [BC] and most probably during this time. The coincidence of this event with a major reorganization of the copper industry of the region—attributed to the results of the campaign of Pharaoh Shoshenq I—raises the possibility that the two were connected, and that camels were introduced as part of the efforts to improve efficiency by facilitating trade.

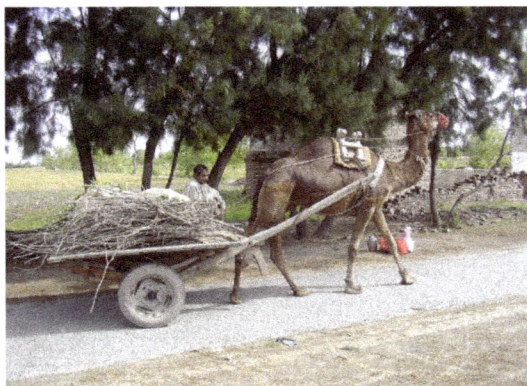

A camel serving as a draft animal in Pakistan (2009).

Petroglyph of a camel, Negev, southern Israel (prior to c. 5300 BC).

## Textiles

Desert tribes and Mongolian nomads use camel hair for tents, yurts, clothing, bedding and accessories. Camels have outer guard hairs and soft inner down, and the fibers are sorted by color and age of the animal. The guard hairs can be felted for use as waterproof coats for the herdsmen, while the softer hair is used for premium goods. The fiber can be spun for use in weaving or made into yarns for hand knitting or crochet. Pure camel hair is recorded as being used for western garments from the 17th century onwards, and from the 19th century a mixture of wool and camel hair was used.

## Military Uses

A special BSF camel contingent, Republic Day Parade, New Delhi.

Camel Corps at Magdhaba, Egypt.

By at least 1200 BC the first camel saddles had appeared, and Bactrian camels could be ridden. The first saddle was positioned to the back of the camel, and control of the Bactrian camel was exercised by means of a stick. However, between 500 and 100 BC, Bactrian camels came into military use. New saddles, which were inflexible and bent, were put over the humps and divided the rider's weight over the animal. In the seventh century BC the military Arabian saddle evolved, which again improved the saddle design slightly.

Military forces have used camel cavalries in wars throughout Africa, the Middle East, and into the modern-day Border Security Force (BSF) of India (though as of July 2012, the BSF planned the replacement of camels with ATVs). The first documented use of camel cavalries occurred in the Battle of Qarqar in 853 BC. Armies have also used camels as freight animals instead of horses and mules.

The East Roman Empire used auxiliary forces known as *dromedarii*, whom the Romans recruited in desert provinces. The camels were used mostly in combat because of their ability to scare off horses at close range (horses are afraid of the camels' scent), a quality famously employed by the Achaemenid Persians when fighting Lydia in the Battle of Thymbra (547 BC).

A camel caravan of the Bulgarian military during the First Balkan War.

The United States Army established the U.S. Camel Corps, stationed in California, in the late 19th century. One may still see stables at the Benicia Arsenal in Benicia, California, where they nowadays serve as the Benicia Historical Museum. Though the experimental use of camels was seen as a success (John B. Floyd, Secretary of War in 1858, recommended that funds be allocated towards obtaining a thousand more camels), the outbreak of the American Civil War in 1861 saw the end of the Camel Corps: Texas became part of the Confederacy, and most of the camels were left to wander away into the desert.

France created a *méhariste* camel corps in 1912 as part of the Armée d'Afrique in the Sahara in order to exercise greater control over the camel-riding Tuareg and Arab insurgents, as previous efforts to defeat them on foot had failed. The Free French Camel Corps fought during World War II, and camel-mounted units remained in service until the end of French rule over Algeria in 1962.

In 1916, the British created the Imperial Camel Corps. It was originally used to fight the Senussi, but was later used in the Sinai and Palestine Campaign in World War I. The Imperial Camel Corps comprised infantrymen mounted on camels for movement across desert, though they dismounted at battle sites and fought on foot. After July 1918, the Corps began to become run down, receiving no new reinforcements, and was formally disbanded in 1919.

In World War I, the British Army also created the Egyptian Camel Transport Corps, which consisted of a group of Egyptian camel drivers and their camels. The Corps supported British war operations in Sinai, Palestine, and Syria by transporting supplies to the troops.

The Somaliland Camel Corps was created by colonial authorities in British Somaliland in 1912; it was disbanded in 1944.

Bactrian camels were used by Romanian forces during World War II in the Caucasian region.

The Bikaner Camel Corps of British India fought alongside the British Indian Army in World Wars I and II.

The *Tropas Nómadas* (Nomad Troops) were an auxiliary regiment of Sahrawi tribesmen serving in the colonial army in Spanish Sahara (today Western Sahara). Operational from the 1930s until the end of the Spanish presence in the territory in 1975, the *Tropas Nómadas* were equipped with small arms and led by Spanish officers. The unit guarded outposts and sometimes conducted patrols on camelback.

## Food Uses

### Dairy

Camels at the Khan and old bridge, Lajjun, Palestine (now in Israel) - 1870s drawing.

A camel calf nursing on camel milk.

Camel milk is a staple food of desert nomad tribes and is sometimes considered a meal itself; a nomad can live on only camel milk for almost a month. Camel milk is rich in vitamins, minerals, proteins, and immunoglobulins; compared to cow's milk, it is lower in fat and lactose, and higher in potassium, iron, and vitamin C. Bedouins believe the curative powers of camel milk are enhanced if the camel's diet consists of certain desert plants. Camel milk can readily be made into a drinkable yogurt, as well as butter or cheese, though the yields for cheese tend to be low.

Camel milk cannot be made into butter by the traditional churning method. It can be made if it is soured first, churned, and a clarifying agent is then added. Until recently, camel milk could not be made into camel cheese because rennet was unable to coagulate the milk proteins to allow the collection of curds. Developing less wasteful uses of the milk, the FAO commissioned Professor J.P. Ramet of the École Nationale Supérieure d'Agronomie et des Industries Alimentaires, who was able to produce curdling by the addition of calcium phosphate and vegetable rennet. The cheese produced from this process has low levels of cholesterol and is easy to digest, even for the lactose intolerant. The sale of camel cheese is limited owing to the small output of the few dairies producing camel cheese and the absence of camel cheese in local (West African) markets. Cheese imports from countries that traditionally breed camels are difficult to obtain due to restrictions on dairy imports from these regions.

Additionally, camel milk can be made into ice cream.

## Meat

A Somali camel meat and rice dish.

Camel meat pulao, from Pakistan.

They provide food in the form of meat and milk. A camel carcass can provide a substantial amount of meat. The male dromedary carcass can weigh 300–400 kg (661–882 lb), while the carcass of a male Bactrian can weigh up to 650 kg (1,433 lb). The carcass of a female dromedary weighs less than the male, ranging between 250 and 350 kg (550 and 770 lb). The brisket, ribs and loin are among the preferred parts, and the hump is considered a delicacy. The hump contains "white and sickly fat", which can be used to make the *khli* (preserved meat) of mutton, beef, or camel. On the other hand, camel milk and meat are rich in protein, vitamins, glycogen, and other nutrients making them essential in the diet of many people. From chemical composition to meat quality, the dromedary camel is the preferred breed for meat production. It does well even in arid areas due to its unusual physiological behaviors and characteristics, which include tolerance to extreme temperatures, radiation from the sun, water paucity, rugged landscape and low vegetation. Camel meat is reported to taste like coarse beef, but older camels can prove to be very tough, although camel meat becomes more tender the more it is cooked. The Abu Dhabi Officers' Club serves a camel burger mixed with beef or lamb fat in order to improve the texture and taste. In Karachi, Pakistan, some restaurants prepare nihari from camel meat. Specialist camel butchers provide expert cuts, with the hump considered the most popular.

Camel meat has been eaten for centuries. It has been recorded by ancient Greek writers as an available dish at banquets in ancient Persia, usually roasted whole. The ancient Roman emperor Heliogabalus enjoyed camel's heel. Camel meat is mainly eaten in certain regions, including Eritrea, Somalia, Djibouti, Saudi Arabia, Egypt, Syria, Libya, Sudan, Ethiopia, Kazakhstan, and other arid regions where alternative forms of protein may be limited or where camel meat has had a long cultural history. Camel blood is also consumable, as is the case among pastoralists in northern Kenya, where camel blood is drunk with milk and acts as a key source of iron, vitamin D, salts and minerals. Camel Meat is a sample food in Djiboutian cuisine. You can have Camel Steak, Skewers and Hamburgers also Zurbyaan rice with camel meat is served in local restaurants. Camel meat is also occasionally found in Australian cuisine: for example, a camel lasagna is available in Alice Springs.

A 2005 report issued jointly by the Saudi Ministry of Health and the United States Centers for Disease Control and Prevention details cases of human bubonic plague resulting from the ingestion of raw camel liver.

# Domestic Rabbit

A domestic or domesticated rabbit (Oryctolagus cuniculus)—more commonly known as a pet rabbit, a bunny, or a bunny rabbit—is a species of European rabbit. A domestic rabbit kept as a pet may be considered a pocket pet, depending on its size. A male rabbit is known as a buck, a female is a doe, and a young rabbit is a kit, or kitten.

Rabbits were first exploited by the Romans as sources of food and fur, and have been kept as pets in Western nations since the 19th century. Beginning in the 1980s, the idea of the domestic rabbit as a house companion, a so-called house rabbit, was promoted. Rabbits can be litter box-trained and may come when called, but they need exercise and can damage a house that is not "rabbit proof." Unwanted rabbits end up in animal shelters, especially after the Easter season. Because they have become invasive in Australia, pet rabbits are banned in the state of Queensland.

## Biology

### Genetics

The study of rabbit genetics is of interest to fanciers, the fiber & fur industry, medical researchers, and the meat industry. Among rabbit fanciers, the genetics of rabbit health and diversity are paramount. The fiber & fur industry focuses on the genetics of coat color and hair properties. In the biomedical research community and the pharmaceutical industry, rabbit genetics are important in model organism research, antibody production, and toxicity testing. The meat industry relies on genetics for disease resistance, feed conversion ratios, and reproduction potential in rabbits.

The rabbit genome has been sequenced and is publicly available. The mitochondrial DNA has also been sequenced. In 2011, parts of the rabbit genome were re-sequenced in greater depth in order to expose variation within the genome.

### Rabbit Coat Pattern and Color Genes

There are 11 color gene groups (or loci) in rabbits. A rabbit's coat has either two pigments (pheomelanin for yellow, and eumelanin for dark brown) or no pigment (for an albino rabbit). Clusters of color genes plus their modifiers control such aspects as coat patterns (e.g. *Dutch* or *English* markings), color hues and their intensity or dilution, and the location of color bands on the hair shaft (e.g., silvering).

Gene = En Pattern: English Gene = A- B- C- D- E- Color: Chestnut.

Gene = Enen Pattern: Broken  Gene = D Color: Chocolate (on white).

Gene = e(j) Pattern: Harlequin.

Gene = du Pattern: Dutch Gene = B Color: Black (on white).

## Diet

As a refinement of the diet of the wild rabbit, the diet of the domestic rabbit is often a function of its purpose. Show rabbits are fed for vibrant health, strong musculoskeletal systems, and—like rabbits intended for the fur trade—optimal coat production and condition. Rabbits intended for the meat trade are fed for swift and efficient production of flesh, while rabbits in research settings have closely controlled diets for specific goals. Nutritional needs of the domestic rabbit may also be focused on developing a physique that allows for the safe delivery of larger litters of healthy kits. Optimizing costs and producing feces that meet local waste regulations may also be factors. The diet of a pet rabbit, too, is geared toward its purpose—as a healthy and long-lived companion.

Hay is an essential part of the diet of all rabbits and it is a major component of the commercial food pellets that are formulated for domestic rabbits and available in many areas. Pellets are typically fed to adult rabbits in limited quantities once or twice a day, to mimic their natural behavior and to prevent obesity. It is recommended only a teaspoon to a egg cup full of pellets is fed to adult rabbits each day. Most rabbit pellets are alfalfa-based for protein and fiber, with other grains completing the carbohydrate requirements. "Muesli" style rabbit foods are also available; these contain separate components--e.g., dried carrot, pea flakes and hay pellets as opposed to a uniform pellet. These are not recommended as rabbits will choose favored parts and leave the rest. Muesli style feeds are often lower in fiber than pelleted versions of rabbit food. Additionally numerous studies have found they increase the risk of obesity and dental disease. Minerals and vitamins are added during production of rabbit pellets to meet the nutritional requirements of the domestic rabbit. Along with pellets, many commercial rabbit raisers also feed one or more types of loose hay, for its freshness and important cellulose components. Alfalfa in particular is recommended for the growth needs of young rabbits.

## Digestion

Rabbits are hindgut fermenters and therefore have an enlarged cecum. This allows a rabbit to digest, via fermentation, what it otherwise would not be able to metabolically process.

After a rabbit ingests food, the food travels down the esophagus and through a small valve called the cardia. In rabbits, this valve is very well pronounced and makes the rabbit incapable of vomiting. The food enters the stomach after passing through the cardia. Food then moves to the stomach and small intestine, where a majority of nutrient extraction and absorption takes place. Food then passes into the colon and eventually into the cecum. Peristaltic muscle contractions (waves of motion) help to separate fibrous and non-fibrous particles. The non-fibrous particles are then moved backwards up the colon, through the illeo-cecal valve, and into the cecum. Symbiotic bacteria in the cecum help to further digest the non-fibrous particles into a more metabolically manageable substance. After as little as three hours, a soft, fecal "pellet," called a cecotrope, is expelled from the rabbit's anus. The rabbit instinctively eats these grape-like pellets, without chewing, in exchange keeping the mucous coating intact. This coating protects the vitamin- and nutrient-rich bacteria from stomach acid, until it reaches the small intestine, where the nutrients from the cecotrope can be absorbed.

The soft pellets contain a sufficiently large portion of nutrients that are critical to the rabbit's health. This soft fecal matter is rich in vitamin B and other nutrients. The process of coprophagy is important to the stability of a rabbit's digestive health because it is one important way that which a rabbit receives vitamin B in a form that is useful to its digestive wellness. Occasionally, the rabbit may leave these pellets lying about its cage; this behavior is harmless and usually related to an ample food supply.

When caecal pellets are wet and runny (semi-liquid) and stick to the rabbit and surrounding objects, they are called intermittent soft cecotropes (ISCs). This is different from ordinary diarrhea and is usually caused by a diet too high in carbohydrates or too low in fiber. Soft fruit or salad items such as lettuce, cucumbers and tomatoes are possible causes.

## Reproduction

Rabbits have a reputation as prolific breeders, and deservedly so, in part because rabbits reach breeding age quickly. To prevent unwanted offspring and to benefit the rabbit's health and behavior, rabbits may be spayed or neutered at sexual maturity: 4–5 months for small breeds (e.g., Mini Rex, Netherland Dwarf), 5–6 months for medium-sized breeds (e.g., Rex, New Zealand), and 6–7 months for large breeds (e.g., Flemish Giant). Bucks usually require more time to sexually mature than does, and they normally reach adult sperm counts at 6–7 months.

Rabbits, like all mammals, produce milk for their young. Female rabbits have six to eight nipples and produce milk for four weeks after birthing. Rabbit milk is relatively high in fat, as a percentage by mass. While most species produce approximately 5% milk fat, rabbits produce 12%. The excerpted table below compares milk characteristics among mammals.

| Species | Fat % | Protein % | Lactose % | Ash % | Total Solids % |
|---------|-------|-----------|-----------|-------|----------------|
| Gray Seal | 53.2 | 11.2 | 2.6 | 0.7 | 67.7 |
| Polar Bear | 31.0 | 10.2 | 0.5 | 1.2 | 42.9 |
| Rabbit | 12.2 | 10.4 | 1.8 | 2.0 | 26.4 |
| Bison | 1.7 | 4.8 | 5.7 | .96 | 13.2 |
| Donkey | 1.2 | 1.7 | 6.9 | .45 | 10.2 |

## Health

Disease is rare when rabbits are raised in sanitary conditions and provided with adequate care. Rabbits have fragile bones, especially in their spines, and need support on the belly or bottom when they are picked up.

Spayed or neutered rabbits kept indoors with proper care may have a lifespan of 8 to 12 years, with mixed-breed rabbits typically living longer than purebred specimens, and dwarf breeds having longer average lifespans than larger breeds. The world record for longest-lived rabbit is 18 years.

Rabbits will gnaw on almost anything, including electrical cords (possibly leading to electrocution), potentially poisonous plants, and material like carpet and fabric that may cause life-threatening intestinal blockages, so areas to which they have access need to be rabbit-proofed.

## Spaying and Neutering

Rabbit fancier organizations and veterinarians recommend that pet rabbits be spayed or neutered by a rabbit-experienced veterinarian. Health advantages of surgically altering a rabbit include increased longevity and (for females) a reduced risk of ovarian and uterine cancers or of endometritis. For both rabbit sexes, spaying or neutering reduces aggression toward other rabbits, as well as territorial marking (especially in males). Rabbits are at high risk for complications from anesthesia and infection of the surgical site is another top concern. Since un-altered animals are not as likely to form agreeable social bonds, spaying and neutering promotes less stressful interactions.

## Vaccinations

In most jurisdictions, including the United States (except where required by local animal control ordinances), rabbits do not require vaccination. Vaccinations exist for both rabbit hemorrhagic disease and myxomatosis. These vaccinations are usually given annually, two weeks apart. If there is an outbreak of myxomatosis locally, this vaccine can be administered every six months for extra protection. Myxomatosis immunizations are not available in all countries, including Australia, due to fears that immunity will pass on to feral rabbits. However, they are recommended by some veterinarians as prophylactics, where they are legally available. In the UK a combined vaccination exists for myxomatosis and VHD1 made by Nobivac called Myxo-RHD, this is given yearly. Due to increasing cases of VHD2 it is now recommended rabbits receive an additional vaccination for RHD2 one brand for this is filovac, the vaccination is given yearly 2 weeks apart from other vaccinations, it may be given 6 monthly at rabbit believed to be at higher risk.

## Declawing

A rabbit cannot be declawed. Lacking pads on the bottoms of its feet, a rabbit requires its claws for traction. Removing its claws would render it unable to stand.

## Tonic Immobility

Coping with stress is a key aspect of rabbit behavior, and this can be traced to part of the brain known as ventral tegmental area (VTA). Dopaminergic neurons in this part of the brain release the hormone dopamine, generalized as a *feel-good* hormone. In humans, dopamine is released through a variety of acts, including sexual activity, substance abuse, and even eating chocolate. However, in rabbits, it is released as part of a coping mechanism while in a heightened state of fear or stress, and has a calming effect. Dopamine has also been found in the rabbit's medial prefrontal cortex, the nucleus accumbens, and the amygdala. Physiological and behavioral responses to human-induced tonic immobility (TI, sometimes termed "trancing" or "playing dead") have been found to be indicative of a fear-motivated stress state, confirming that the promotion of TI to try to increase a bond between rabbits and their owners—thinking the rabbits enjoy it—is misplaced. However, some researchers conclude that inducing TI in rabbits is appropriate for certain procedures, as it holds less risk than anesthesia.

## Sore Hocks

The formation of open sores on the rabbit's hocks, commonly called *sore hocks*, is a problem that commonly afflicts mostly heavy-weight rabbits kept in cages with wire flooring or soiled solid flooring. The problem is most prevalent in rex-furred rabbits and heavy-weight rabbits (over 9 pounds (4.1 kg)), as well as those with thin foot bristles.

The condition results when, over the course of time, the protective bristle-like fur on the rabbit's hocks thins down. Standing urine or other unsanitary cage conditions can exacerbate the problem by irritating the sensitive skin. The exposed skin in turn can result in tender areas or, in severe cases, open sores, which may then become infected and abscessed if not properly cared for.

## Gastrointestinal Stasis

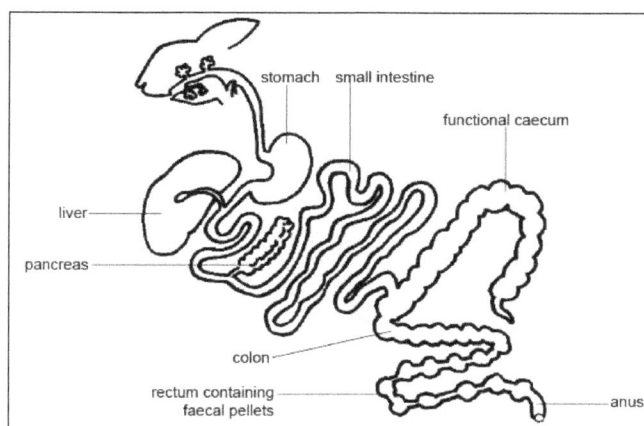

Digestive system of the rabbit.

Gastrointestinal stasis (GI stasis) is a serious and potentially fatal condition that occurs in some rabbits in which gut motility is severely reduced and possibly completely stopped. When untreated or improperly treated, GI stasis can be fatal in as little as 24 hours.

GI stasis is the condition of food not moving through the gut as quickly as normal. The gut contents may dehydrate and compact into a hard, immobile mass (impacted gut), blocking the digestive tract of the rabbit. Food in an immobile gut may also ferment, causing significant gas buildup and resultant gas pain for the rabbit.

The first noticeable symptom of GI stasis may be that the rabbit suddenly stops eating. Treatment frequently includes intravenous or subcutaneous fluid therapy (rehydration through injection of a balanced electrolyte solution), pain control, possible careful massage to promote gas expulsion and comfort, drugs to promote gut motility, and careful monitoring of all inputs and outputs. The rabbit's diet may also be changed as part of treatment, to include force-feeding to ensure adequate nutrition. Surgery to remove the blockage is not generally recommended and comes with a poor prognosis.

Some rabbits are more prone to GI stasis than others. The causes of GI stasis are not completely understood, but common contributing factors are thought to include stress, reduced food intake, low fiber in the diet, dehydration, reduction in exercise or blockage caused by excess fur or carpet ingestion. Stress factors can include changes in housing, transportation, or medical procedures under anesthesia. As many of these factors may occur together (poor dental structure leading to decreased food intake, followed by a stressful veterinary dental procedure to correct the dental problem) establishing a root cause may be difficult.

GI stasis is sometimes misdiagnosed as "hair balls" by veterinarians or rabbit keepers not familiar with the condition. While fur is commonly found in the stomach following a fatal case of GI stasis, it is also found in healthy rabbits. Molting and chewing fur can be a predisposing factor in the occurrence of GI stasis, however, the primary cause is the change in motility of the gut.

## Dental Problems

Dental disease has several causes, namely genetics, inappropriate diet, injury to the jaw, infection, or cancer.

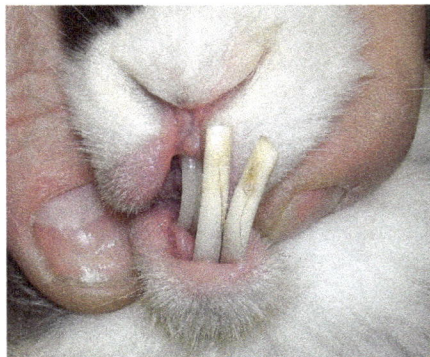

Malocclusion in a rabbit.

- Malocclusion: Rabbit teeth are open-rooted and continue to grow throughout their lives. In some rabbits, the teeth are not properly aligned, a condition called *malocclusion*. Because of the misaligned nature of the rabbit's teeth, there is no normal wear to control the length to which the teeth grow. There are three main causes of malocclusion, most commonly genetic predisposition, injury, or bacterial infection. In the case of congenital malocclusion, treatment usually involves veterinary visits in which the teeth are treated with a dental burr (a procedure called crown reduction or, more commonly, teeth clipping) or, in some cases, permanently removed. In cases of simple malocclusion, a block of wood for the rabbit to chew on can rectify this problem.

- Molar spurs: These are spurs that can dig into the rabbit's tongue and/or cheek causing pain. These should be filed down by an experienced exotic veterinarian specialised in rabbit care, using a dental burr.

- Osteoporosis: Rabbits, especially neutered females and those that are kept indoors without adequate natural sunlight, can suffer from osteoporosis, in which holes appear in the skull by X-Ray imaging. This reflects the general thinning of the bone, and teeth will start to become looser in the sockets, making it uncomfortable and painful for the animal to chew hay. The inability to properly chew hay can result in molar spurs, as described above, and weight loss, leading into a downward spiral if not treated promptly. This can be reversible and treatable. A veterinary formulated liquid calcium supplement with vitamin D3 and magnesium can be given mixed with the rabbit's drinking water, once or twice per week, according to the veterinarian's instructions. The molar spurs should also be trimmed down by an experienced exotic veterinarian specialised in rabbit care, once per 1–2 months depending on the case.

Signs of dental difficulty include difficulty eating, weight loss and small stools and visibly overgrown teeth. However, there are many other causes of ptyalism, including pain due to other causes.

## Respiratory and Conjunctival Problems

An over-diagnosed ailment amongst rabbits is respiratory infection, known colloquially as "snuffles". *Pasteurella*, a bacterium, is usually misdiagnosed and this is known to be a factor in the overuse of antibiotics among rabbits. A runny nose, for instance, can have several causes, among those being high temperature or humidity, extreme stress, environmental pollution (like perfume or incense), or a sinus infection. Options for treating this is removing the pollutant, lowering or raising the temperature accordingly, and medical treatment for sinus infections. *Pasteurella* does live naturally in a rabbit's respiratory tract, and it can flourish out of control in some cases. In the rare event that happens, antibiotic treatment is necessary.

Sneezing can be a sign of environmental pollution (such as too much dust) or a food allergy.

Runny eyes and other conjunctival problems can be caused by dental disease or a blockage of the tear duct. Environmental pollution, corneal disease, entropion, distichiasis, or inflammation of the eyes are also causes. This is easy to diagnose as well as treat.

## Viral Diseases

Rabbits are subject to infection by a variety of viruses. Some have had deadly and widespread impact.

Myxomatosis Trial.

## Myxomatosis

Myxomatosis is a virulent threat to all rabbits but not to humans. The intentional introduction of myxomatosis in rabbit-ravaged Australia killed an estimated 500 million feral rabbits between

1950 and 1952. The Australian government will not allow veterinarians to purchase and use the myxomatosis vaccine that would protect domestic rabbits, for fear that this immunity would be spread into the wild via escaped livestock and pets. This potential consequence is also one motivation for the pet-rabbit ban in Queensland.

In Australia, rabbits caged outdoors in areas with high numbers of mosquitoes are vulnerable to myxomatosis. In Europe, fleas are the carriers of myxomatosis. In some countries, annual vaccinations against myxomatosis are available.

## Rabbit Hemorrhagic Disease (RHD)

Rabbit hemorrhagic disease (RHD), also known as *viral hemorrhagic disease* (VHD) or *rabbit calicivirus disease* (RCD), is caused by a rabbit-specific calicivirus known as RHDV or RCV. Discovered in 1983, RHD is highly infectious and usually fatal. Initial signs of the disease may be limited to fever and lethargy, until significant internal organ damage results in labored breathing, squealing, bloody mucus, and eventual coma and death. Internally, the infection causes necrosis of the liver and damages other organs, especially the spleen, kidneys, and small intestine.

RHD, like myxomatosis, has been intentionally introduced to control feral rabbit populations in Australia and (illegally) in New Zealand, and RHD has, in some areas, escaped quarantine. The disease has killed tens of millions of rabbits in China (unintentionally) as well as Australia, with other epidemics reported in Bolivia, Mexico, South Korea, and continental Europe. Rabbit populations in New Zealand have bounced back after developing a genetic immunity to RHD, and the disease has, so far, had no effect on the genetically divergent native wild rabbits and hares in the Americas.

In the United States, an October 2013 USDA document stated:

> "RHD has been found in the United States as recently as 2010, and was detected in Canada in 2011. Thus far, outbreaks have been controlled quickly through quarantine, depopulation, disease tracing, and cleaning and disinfection; however, rabbit losses have been in the thousands. An RHD vaccine exists, but it is not recommended for use where the disease is not widespread in wildlife, as it may hide signs of disease and is not considered a practical response for such a rapidly spreading disease".

In the UK, reports of RHD (as recently as February 2018) have been submitted to the British Rabbit Council's online "Notice Board". Vaccines for RHD are available—and mandatory—in the UK.

## West Nile Virus

West Nile virus is another threat to domestic as well as wild rabbits. It is a fatal disease, and while vaccines are available for other species, there are none yet specifically indicated for rabbits.

## Wry Neck and Parasitic Fungus

Wry neck (or head tilt) is a condition in rabbits that can be fatal, due to the resulting disorientation that causes the animal to stop eating and drinking. Inner ear infections or ear mites, as well as diseases or injuries affecting the brain (including stroke) can lead to wry neck. The most common cause, however, is a parasitic microscopic fungus called Encephalitozoon cuniculi (E. cuniculi).

"Despite approximately half of all pet rabbits carrying the infection, only a small proportion of these cases ever show any illness". Some vets now recommend treating rabbits against E. cuniculi. The usual drugs for treatment and prevention are the benzimidazole anthelmintics, particularly fenbendazole (also used as a deworming agent in other animal species). In the UK, fenbendazole (under the brand name Panacur Rabbit), is sold over-the-counter in oral paste form as a nine-day treatment. Fenbendazole is particularly recommended for rabbits kept in colonies and as a preventive before mixing new rabbits with each other.

## Fly Strike

Fly strike, or blowfly strike, (*Lucilia sericata*) is a condition that occurs when flies (particularly botflies) lay their eggs in a rabbit's damp or soiled fur, or in an open wound. Within 12 hours, the eggs hatch into the larval stage of the fly, known as maggots. Initially small but quickly growing to 15 millimetres (0.59 in) long, maggots can burrow into skin and feed on an animal's tissue, leading to shock and death. The most susceptible rabbits are those in unsanitary conditions, sedentary ones, and those unable to clean their excretory areas. Rabbits with diarrhea should be inspected for fly strike, especially during the summer months. The topical treatment Rearguard (from Novartis) is approved in the United Kingdom for 10-week-per-application prevention of fly strike.

## Breeds

Gemüsestilleben mit Häschen, by Johann Georg Seitz.

As of 2017, there were at least 305 breeds of domestic rabbit in 70 countries around the world. The American Rabbit Breeders Association currently recognizes 49 rabbit breeds and the British Rabbit Council recognizes 106. Selective breeding has produced rabbits ranging in size from dwarf to giant. Across the world, rabbits are raised as livestock (in *cuniculture)* for their meat, pelts, and wool, and also by fanciers and hobbyists as pets.

Rabbits have been selectively bred since ancient times to achieve certain desired characteristics. Variations include size and body shape, coat type (including hair length and texture), coat color, ear carriage (erect or lop), and even ear length. As with any animal, domesticated rabbits' temperaments vary in such factors as energy level and novelty seeking.

Most genetic defects in the domestic rabbit (such as dental problems in the Holland Lop breed) are due to recessive genes. Genetics are carefully tracked by fanciers who show rabbits, to breed out defects.

## As Livestock

Rabbits have been kept as livestock since ancient times for their meat, wool, and fur. In modern times, rabbits are also utilized in scientific research as laboratory animals.

## Meat Rabbits

Meat-breed rabbits were a supplementary food
source during the Great Depression.

Breeds such as the New Zealand and Californian are frequently utilized for meat in commercial rabbitries. These breeds have efficient metabolisms and grow quickly; they are ready for slaughter by approximately 14 to 16 weeks of age.

Rabbit fryers are rabbits that are between 70 and 90 days of age, and weighing between 3 and 5 lb (1 to 2 kg) live weight. Rabbit roasters are rabbits from 90 days to 6 months of age weighing between 5 and 8 lb (2 to 3.5 kg) live weight. Rabbit stewers are rabbits from 6 months on weighing over 8 lb.

Any type of rabbit can be slaughtered for meat, but those exhibiting the "commercial" body type are most commonly raised for meat purposes. Dark fryers (any other color but albino whites) are sometimes lower in price than albino fryers because of the slightly darker tinge of the fryer (purely pink carcasses are preferred by consumers) and because the dark hairs are easier to see than if there are residual white hairs on the carcass. There is no difference in skinability.

## Wool Rabbits

Rabbits such as the Angora, American Fuzzy Lop, and Jersey Wooly produce wool. However, since the American Fuzzy Lop and Jersey Wooly are both dwarf breeds, only the much larger Angora breeds such as the English Angora, Satin Angora, Giant Angora, and French Angoras are used for commercial wool production. Their long fur is sheared, combed, or plucked (gently pulling loose hairs from the body during molting) and then spun into yarn used to make a variety of products. Angora sweaters can be purchased in many clothing stores and is generally mixed with other types of wool. Rabbit wool, called Angora, is 2.5 times warmer than sheep's wool.

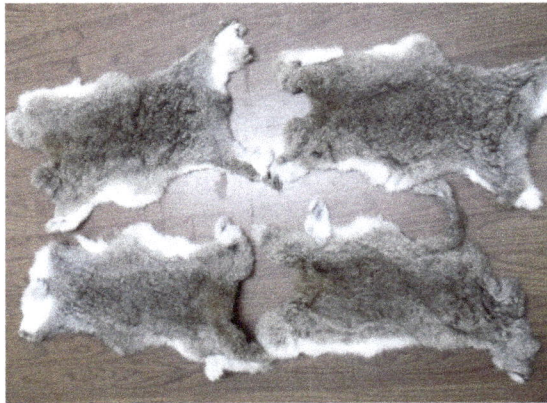

Peaux de Lapin ("Rabbit skins") by Edme Bouchardon.

## Fur Rabbits

Rabbit breeds that were developed for their fur qualities include the Rex with its plush texture, the Satin with its lustrous color, and the Chinchilla for its exotic pattern. White rabbit fur may be dyed in an array of colors that aren't produced naturally. Rabbits in the fur industry are fed a diet focused for robust coat production and pelts are harvested after the rabbit reaches prime condition, which takes longer than in the meat industry. Rabbit fur is used in local and commercial textile industries throughout the world. China imports much of its rabbit fur from Scandinavia (80%) and some from North America (5%).

## Laboratory Rabbits

Rabbits have been and continue to be used in laboratory work such as production of antibodies for vaccines and research of human male reproductive system toxicology. In 1972, around 450,000 rabbits were used for experiments in the United States, decreasing to around 240,000 in 2006. The Environmental Health Perspective, published by the National Institute of Health, states, "The rabbit is an extremely valuable model for studying the effects of chemicals or other stimuli on the male reproductive system." According to the Humane Society of the United States, rabbits are also used extensively in the study of bronchial asthma, stroke prevention treatments, cystic fibrosis, diabetes, and cancer.

The New Zealand White is one of the most commonly used breeds for research and testing.

Pasture-raised rabbits in a moveable enclosure at Polyface Farm.

Animal rights activists generally oppose animal experimentation for all purposes, and rabbits are no exception. The use of rabbits for the Draize test, which is used for, amongst other things, testing cosmetics on animals, has been cited as an example of cruelty in animal research. Albino rabbits are typically used in the Draize tests because they have less tear flow than other animals, and the lack of eye pigment makes the effects easier to visualize.

## Housing

Rabbits can live outdoors in properly constructed, sheltered hutches, which provide protection from the elements in winter and keep rabbits cool in summer heat. To protect from predators, rabbit hutches are usually situated in a fenced yard, shed, barn, or other enclosed structure, which may also contain a larger pen for exercise. Rabbits in such an environment can alternatively be allowed to roam the secured area freely, and simply be provided with an adapted doghouse for shelter. A more elaborate setup is an artificial warren.

# Horses

Horse is a hoofed herbivorous mammal of the family Equidae. It comprises a single species, Equus caballus, whose numerous varieties are called breeds. Before the advent of mechanized vehicles, the horse was widely used as a draft animal, and riding on horseback was one of the chief means of transportation.

## General Features

In prehistoric times the wild horse was probably first hunted for food. Research suggests that domestication had taken place by approximately 5,500 years ago. It is supposed that the horse was first used by a tribe of Indo-European origin that lived in the steppes north of the chain of mountains adjacent to the Black and Caspian seas. Influenced by climate, food, and humans, the horse rapidly acquired its present form.

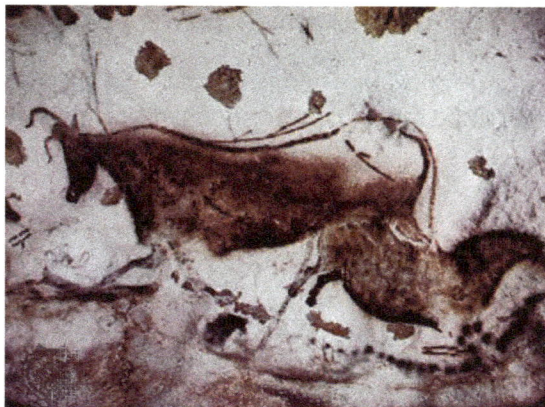

Cave painting of a bull and a horse; in Lascaux Grotto, near Montignac, France.

The relationship of the horse to humans has been unique. The horse is a partner and friend. It has plowed fields and brought in the harvest, hauled goods and conveyed passengers, followed game and tracked cattle, and carried combatants into battle and adventurers to unknown lands. It has provided recreation in the form of jousts, tournaments, carousels, and the sport of riding. The influence of the horse is expressed in the English language in such terms as chivalry and cavalier, which connote honour, respect, good manners, and straightforwardness.

A team of Clydesdales pulling a plow at a draft horse demonstration.

The horse is the "proudest conquest of Man," according to the French zoologist Georges-Louis Leclerc, comte de Buffon. Its place was at its master's side in the graves of the Scythian kings or in the tombs of the pharaohs. Many early human cultures were centred on possession of the horse. Superstition read meaning into the colours of the horse, and a horse's head suspended near a grave or sanctuary or on the gables of a house conferred supernatural powers on the place. Greek mythology created the Centaur, the most obvious symbol of the oneness of horse and rider. White stallions were the supreme sacrifice to the gods, and the Greek general Xenophon recorded that "gods and heroes are depicted on well-trained horses." A beautiful and well-trained horse was, therefore, a status symbol in ancient Greece. Kings, generals, and statesmen, of necessity, had to be horsemen. The names of famous horses are inseparably linked to those of their famous riders: Bucephalus, the charger of Alexander the Great; Incitatus, once believed to have been made a senator by the Roman emperor Caligula; El Morzillo, Hernán Cortés's favourite horse, to whom the Indians erected a statue; Roan Barbery, the stallion of Richard II, mentioned by Shakespeare; Copenhagen, the duke of Wellington's horse, which was buried with military honours.

Mosaic of Alexander the Great discovered in the House of the Faun, Pompeii, Italy.

The horse has occupied a special place in the realm of art. From Stone Age drawings to the marvel of the Parthenon frieze, from Chinese Tang dynasty tomb sculptures to Leonardo da Vinci's sketches and Andrea del Verrocchio's Colleoni, from the Qur'ān to modern literature, the horse has inspired artists of all ages and in all parts of the world.

Jade horse head, Chinese, Han dynasty (206 BCE–220 CE); in the Victoria and Albert Museum, London. Height 19 cm.

The horse in life has served its master in travels, wars, and labours and in death has provided many commodities. Long before their domestication, horses were hunted by primitive tribes for their flesh, and horsemeat is still consumed by people in parts of Europe and in Iceland and is the basis of many pet foods. Horse bones and cartilage are used to make glue. Tetanus antitoxin is obtained from the blood serum of horses previously inoculated with tetanus toxoid. From horsehide a number of articles are manufactured, including fine shoes and belts. The cordovan leather fabricated by the Moors in Córdoba, Spain, was originally made from horsehide. Stylish fur coats are made of the sleek coats of foals. Horsehair has wide use in upholstery, mattresses, and stiff lining for coats and suits; high-quality horsehair, usually white, is employed for violin bows. Horse manure, which today provides the basis for cultivation of mushrooms, was used by the Scythians for fuel. Mare's milk was drunk by the Scythians, the Mongols, and the Arabs.

## Form and Function

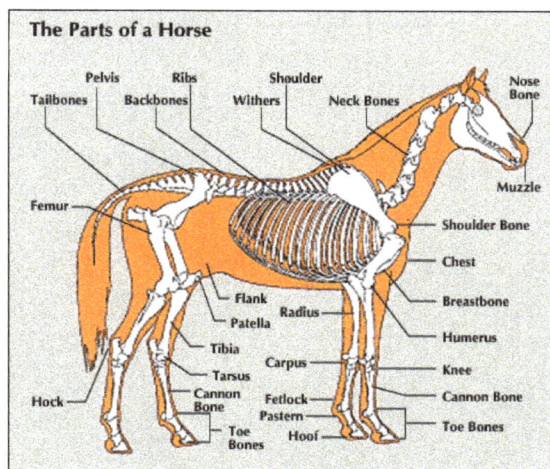

Horse anatomy.

A mature male horse is called a stallion, the female a mare. A stallion used for breeding is known as a stud. A castrated stallion is commonly called a gelding. Formerly, stallions were employed as riding horses, while mares were kept for breeding purposes only. Geldings were used for work and as ladies' riding horses. Recently, however, geldings generally have replaced stallions as riding horses. Young horses are known as foals; male foals are called colts and females fillies.

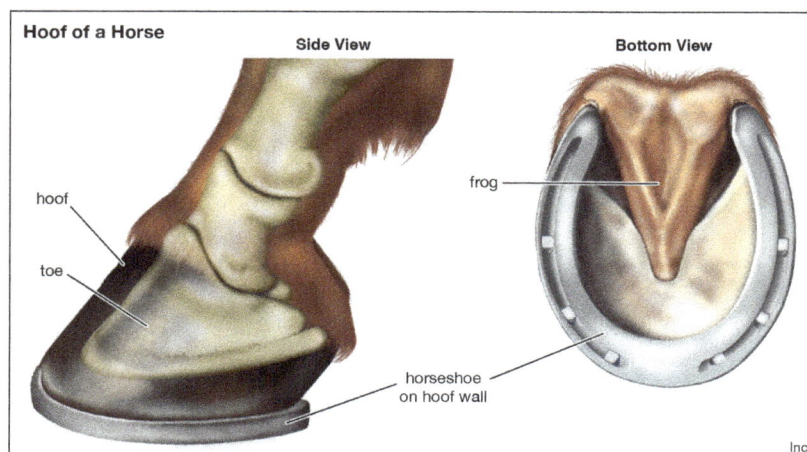

Horse: Hoof of a horse.

## Anatomical Adaptations

The primitive horse probably stood 12 hands (about 120 cm, or 48 inches [1 hand = 10 cm, or 4 inches]) tall at the withers, the high point on the back at the base of the neck, and was dun coloured (typically brownish to dark gray). Domestic horses gone wild, such as the mustangs of western North America, tend to revert to those primitive features under random mating: they generally are somewhat taller (about 15 hands [152.4 cm, or 60 inches]), are usually gray, dun, or brownish in colour, and move in herds led by a stallion.

The horse's general form is characteristic of an animal of speed: the long leg bones pivot on pulley-like joints that restrict movement to the fore and aft, the limbs are levered to muscle masses in such a way as to provide the most efficient use of energy, and the compact body is supported permanently on the tips of the toes, allowing fuller extension of the limbs in running.

The rounded skull houses a large and complex brain, well developed in those areas that direct muscle coordination. While the horse is intelligent among subhuman animals, it is safe to say that the horse is more concerned with the functioning of its acute sensory reception and its musculature than with mental processes. Though much has been written about "educated" horses that appear to exhibit an ability to spell and count, it is generally agreed that in such cases a very perceptive animal is responding to cues from its master. But this ability is remarkable enough in its own right, for the cues are often given unconsciously by the human trainer, and detection of such subtle signals requires extremely sharp perception.

The horse, like other grazing herbivores, has typical adaptations for plant eating: a set of strong, high-crowned teeth, suited to grinding grasses and other harsh vegetation, and a relatively long digestive tract, most of which is intestine concerned with digesting cellulose matter from vegetation. Young horses have milk (or baby) teeth, which they begin to shed at about age two and a half. The permanent teeth, numbering 36 to 40, are completely developed by age four to five years. In the stallion these teeth are arranged as follows on the upper and lower jaws: 12 incisors that cut and pull at grasses; 4 canines, remnants without function in the modern horse and usually not found in mares; 12 premolars and 12 molars, high prisms that continue to grow out of the jaw in order to replace the surfaces worn off in grinding food.

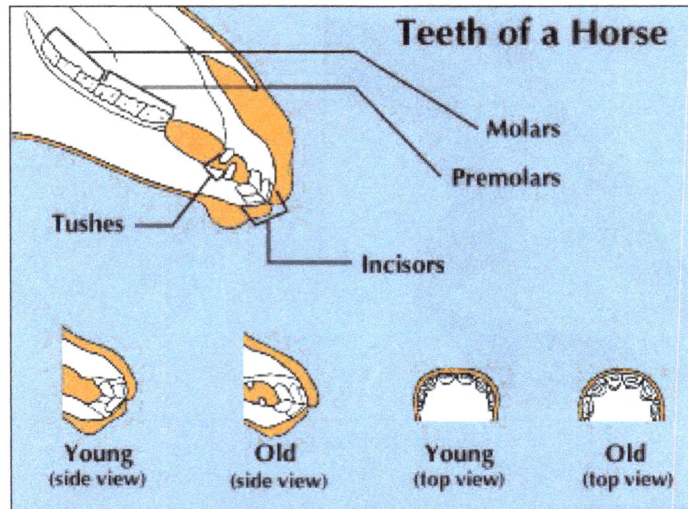

Teeth of a Horse.

Under domestication the horse has diversified into three major types, based on size and build: draft horses, heavy-limbed and up to 20 hands (200 cm, or 80 inches) high; ponies, by convention horses under 14.2 hands (about 147 cm, or 58 inches) high; and light horses—the saddle or riding horses—which fall in the intermediate size range. Domestic horses tend to be nearsighted, less hardy than their ancestors, and often high-strung, especially Thoroughbreds, where intensive breeding has been focused upon speed to the exclusion of other qualities. The stomach is relatively small, and, since much vegetation must be ingested to maintain vital processes, foraging is almost constant under natural conditions. Domestic animals are fed several (at least three) times a day in quantities governed by the exertion of the horse.

## Senses

The extremely large eyes placed far back on the elongated head admirably suit the horse for its chief mode of defense: flight. Its long neck and high-set eyes, which register a much wider range than do the eyes of a human being, enable the horse to discern a possible threat even while eating low grasses. Like human vision, the horse's vision is binocular, but only in the narrow area directly forward. Evidence suggests that a horse's vision is limited in its ability to register colour; horses can detect yellow and blue but not red and green. While visual acuity is high, the eyes do not have variable focus, and objects at different distances register only on different areas of the retina, which requires tilting movements of the head. The senses of smell and hearing seem to be keener than in human beings. As the biologist George Gaylord Simpson put it in Horses:

> "Legs for running and eyes for warning have enabled horses to survive through the ages, although subject to constant attack by flesh eaters that liked nothing better than horse for supper".

## Colour and Pattern

From the dun of the primitive horse has sprung a variety of colours and patterns, some highly variable and difficult to distinguish. Among the most important colours are black, bay, chestnut (and sorrel), palomino, cream, and white.

Common horse colours: dappled gray (top left), dun (centre left), brown (bottom left), strawberry roan (top centre), chestnut (centre), skewbald (a type of pinto, bottom centre), palomino (top right), bay (centre right), black (bottom right).

The black colour is a true black, although a white face marking (blaze) and white ankles (stockings) may occur. The brown horse is almost black but has lighter areas around the muzzle, eyes, and legs. Bay refers to several shades of brown, from red-brown and tan to sandy. Bay horses have a black mane, tail, and (usually) stockings. There is a dilution (or lightening) gene—called silver or silver dapple—that mainly influences the dark colours of the coat. Chestnut is similar to bay but with none of the bay's black overtones. Lighter shades of chestnut are called sorrel. The palomino horse runs from cream to bronze, with a flaxen or silvery mane and tail. The cream is a diluted sorrel, or very pale yellow, nearly white. White in horses is variable, ranging from aging grays to albinos with blue eyes and pink skin and to pseudoalbinos with a buff mane or with brown eyes. The chief patterns of the white horse are gray, roan, pinto, sabino, and appaloosa. Gray horses are born dark brown or black and develop white hairs as they age, becoming almost all white in advanced years. Roan refers to white mixed with other colours at birth: blue roan is white mixed with black; red roan is mixed white and bay; and strawberry roan is white and chestnut. The pinto is almost any spotted pattern of white and another colour; other names, such as paint, calico, piebald, skewbald, overo, and tobiano, refer to subtle distinctions in type of colour or pattern. Appaloosa (leopard complex) is another extremely variable pattern, but the term generally refers to a large white patch over the hips and loin, with scattered irregular dark spots.

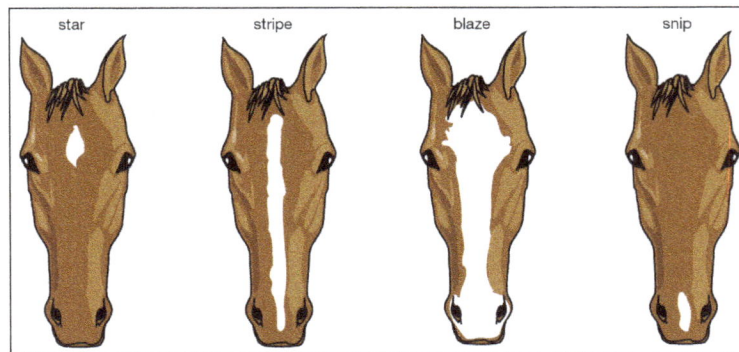

Horse: facial markings
Different types of horse facial markings.

Studies of five coat-colour genes in DNA samples from ancient, predomesticated horses have shown that these horses predominantly carried the genes for black or bay. Scientists believe that it is very likely that these horses also carried the dun dilution gene. The leopard (Appaloosa) mutation was also discovered, which was found to be consistent with some cave paintings dating to 25,000 years ago that depict spotted horses. Mutations for chestnut, tobiano, and sabino were also observed and were dated to 3,000 years ago, whereas the buckskin variant had emerged by about 1,000 years ago. Most of the variation in coat colour appeared after domestication occurred and was likely the result of artificial selection by humans.

Pinto with tobiano pattern.

## Nutrition

The horse's natural food is grass. For stabled horses, the diet generally consists of hay and grain. The animal should not be fed immediately before or after work, to avoid digestive problems. Fresh water is important, especially when the horse is shedding its winter coat, but the animal should never be watered when it is overheated after working. Oats provide the greatest nutritional value and are given especially to foals. Older horses, whose teeth are worn down, or those with digestive troubles, can be provided with crushed oats. Chaff (minced straw) can be added to the oat ration of animals that eat greedily or do not chew the grain properly. Crushed barley is sometimes substituted in part for oats. Hay provides the bulk of the horse's ration and may be of varying composition according to locale. Mash is bran mixed with water and with various invigorating additions or medications. It may be given to horses with digestive troubles or deficient eating habits. Corn (maize) is used as a fattening cereal, but it makes the horse sweat easily. Salt is needed by the horse at all times and especially when shedding. Bread, carrots, and sugar are tidbits often used by the rider or trainer to reward an animal. In times of poverty, horses have adapted to all sorts of food—potatoes, beans, green leaves, and in Iceland even fish—but such foods are not generally taken if other fare is available. A number of commercial feed mixes are available to modern breeders and owners; these mixes contain minerals, vitamins, and other nutrients and are designed to provide a balanced diet when supplemented with hay.

## Behavior

The horse's nervous system is highly developed and gives proof to varying degrees of the essential faculties that are the basis of intelligence: instinct, memory, and judgment. Foals, which stand on their feet a short while after birth and are able to follow their mothers within a few hours, even at this early stage in life exhibit the traits generally ascribed to horses. They have a tendency to flee danger. They express fear sometimes by showing panic and sometimes by immobility. Horses rarely attack and do so either when flight is impossible or when driven to assault a person who has treated them brutally.

Habit governs a large number of their reactions. Instinct, together with a fine sense of smell and hearing, enables them to sense water, fire, even distant danger. An extremely well-developed sense of direction permits the horse to find its way back to its stables even at night or after a prolonged absence. The visual memory of the horse prompts it to shy repeatedly from an object or place where it had earlier experienced fear. The animal's auditory memory, which enabled ancient army horses or hunters to follow the sounds of the bugles, is used in training. When teaching, the instructor always uses the same words and the same tone of voice for a given desired reaction. Intelligent horses soon attach certain movements desired by their trainers to particular sounds and even try to anticipate their rider's wishes.

While instinct is an unconscious reaction more or less present in all individuals of the same species, the degree of its expression varies according to the individual and its development. Most horses can sense a rider's uncertainty, nervousness, or fear and are thereby encouraged to disregard or even deliberately disobey the rider.

Cunning animals have been known to employ their intelligence and physical skill to a determined end, such as opening the latch of a stall or the lid of a chest of oats.

## Reproduction and Development

The onset of adult sex characteristics generally begins at the age of 16 to 18 months. The horse is considered mature, depending on the breed, at approximately three years and adult at five. Fecundity varies according to the breed and may last beyond age 20 with Thoroughbreds and to 12 or 15 with other horses. The gestation period is 11 months; 280 days is the minimum in which the foal can be born with expectation to live. As a rule, a mare produces one foal per mating, twins occasionally, and triplets rarely. The foal is weaned at six months.

The useful life of a horse varies according to the amount of work it is required to do and the maintenance furnished by its owner. A horse that is trained carefully and slowly and is given the necessary time for development may be expected to serve to an older age than a horse that is rushed in its training. Racehorses that enter into races at the age of two rarely remain on the turf beyond eight. Well-kept riding horses, on the contrary, may be used more than 20 years.

The life span of a horse is calculated at six to seven times the time necessary for its physical and mental development—that is, 30 to 35 years at the utmost, the rule being about 20 to 25 years. Ponies generally live longer than larger horses. There are a number of examples of horses that have passed the usual limit of age. The veterinary university of Vienna conserves the skeleton of a Thoroughbred mare of 44 years of age. There have been reports made of horses living to their early 60s in age.

## Diseases and Parasites

Horses are subjected to a number of contagious diseases, such as influenza, strangles, glanders, equine encephalomyelitis, and equine infectious anemia (swamp fever). Their skin is affected by parasites, including certain mites, ticks, and lice. Those with sensitive skin are especially subject to eczemas and abscesses, which may result from neglect or contamination. Sores caused by injuries to the skin from ill-fitting or unclean saddles and bridles are common ailments. The horse's digestive tract is particularly sensitive to spoiled feed, which causes acute or chronic indigestion, especially in hot weather. Worms can develop in the intestine and include the larvae of the botfly, pinworms, tapeworms, and roundworms (ascarids). Overwork and neglect may predispose the horse to pneumonia and rheumatism. The ailment known as roaring is an infection of the larynx that makes the horse inhale noisily; a milder form causes the horse to whistle. Chronic asthma, or "broken wind," is an ailment that is all but incurable. A horse's legs and feet are sensitive to blows, sprains, and overwork, especially if the horse is young or is worked on hard surfaces. Lameness may be caused by bony growths, such as splints, spavins, and ringbones, by soft-tissue enlargements, known as windgalls, thoroughpins, and shoe boils, and by injury to the hooves, including sand crack, split hoof, tread thrush, and acute or chronic laminitis.

## Breeds of Horses

The first intensively domesticated horses were developed in Central Asia. They were small, lightweight, and stocky. In time, two general groups of horses emerged: the southerly Arab-Barb types (from the Barbary coast) and the northerly, so-called cold-blooded types. When, where, and how these horses appeared is disputed. Nevertheless, all modern breeds—the light, fast, spirited breeds typified by the modern Arabian, the heavier, slower, and calmer working breeds typified by the Belgian, and the intermediate breeds typified by the Thoroughbred—may be classified according to where they originated (e.g., Percheron, Clydesdale, and Arabian), by the principal use of the horse (riding, draft, coach horse), and by their outward appearance and size (light, heavy, pony).

## Light Horses

### Arabian

Its long history is obscured by legend, but the Arabian breed, prized for its stamina, intelligence, and character, is known to have been developed in Arabia by the 7th century CE. It is a compact horse with a small head, protruding eyes, wide nostrils, marked withers, and a short back. It usually has only 23 vertebrae, while 24 is the usual number for other breeds. (Variation in vertebrae number is found in a wide diversity of breeds.) Its legs are strong with fine hooves. The coat, tail, and mane are of fine silky hair. While many colours are possible in the breed, gray prevails. The most-famous stud farm is in the region of Najd, Saudi Arabia, but many fine Arabian horses are bred in the United States.

## Thoroughbred

The history of the English Thoroughbred is a long one. Records indicate that a stock of Arab and Barb horses was introduced into England as early as the 3rd century. Conditions of climate, soil, and water favoured development, and selective breeding was long encouraged by those interested in racing. Under the reigns of James I and Charles I, 43 mares, the Royal Mares, were imported into England, and a record, the General Stud Book, was begun in which are inscribed only those horses that may be traced back to the Royal Mares in direct line or to only three other horses imported to England—the Byerly Turk (imported in 1689), the Darley Arabian (after 1700), and the Godolphin Barb (also known as the Godolphin Arabian, imported about 1730). The English Thoroughbred has since been introduced to most countries, where it is bred for racing or used to improve local breeds. The Thoroughbred has a small fine head, a deep chest, and a straight back. Its legs have short bones that allow a long easy stride, and its coat is generally bay or chestnut, rarely black or gray.

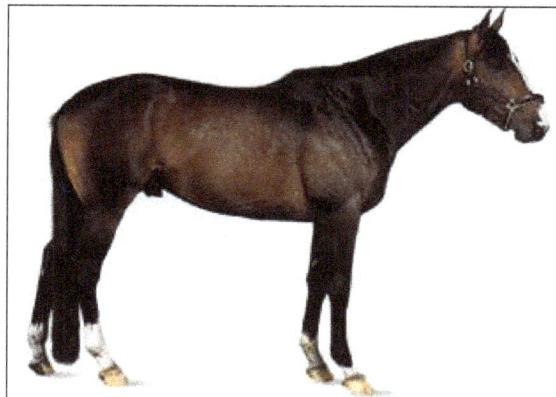

Horse: Thoroughbred stallion with dark bay coat.

## Turkoman, Akhal-teke and Others

Asian breeds were strongly influenced by Arabian or Persian breeds, which together with the horses of the steppes produced small plain-looking horses of great intelligence and endurance. Among them are the Turkoman, Akhal-Teke, Tartar, Kyrgyz, Mongol, and Cossack horses. A Persian stallion and a Dutch mare produced the Orlov trotter in 1778, named after Aleksey Grigoryevich, Count Orlov, the owner of the stud farm in Khrenovoye, Russia, where the mating took place.

## Anglo-arab

The Anglo-Arab breed originated in France with a crossing of English Thoroughbreds with pure Arabians. The matings produced a horse larger than the Arabian and smaller than the Thoroughbred, of easy maintenance, and capable of carrying considerable weight in the saddle. Its coat is generally chestnut or bay.

## American Breeds

The Standardbred, a breed that excels at the pace and trot, ranks as one of the world's finest harness racers. A powerful long-bodied horse, the Standardbred was developed during the first half of the 19th century and can be traced largely to the sire Messenger, a Thoroughbred imported from Britain in 1788 and mated to various brood mares in New York, New Jersey, and Pennsylvania.

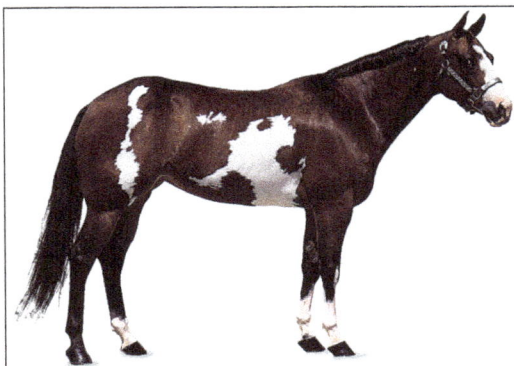

American Paint Horse mare of bay colouring.

The American Quarter Horse was bred for races of a quarter of a mile and is said to descend from Janus, a small Thoroughbred stallion imported into Virginia toward the end of the 18th century. It is 14.2 to 16 hands (about 147 to 162.6 cm, or 58 to 64 inches) high, with sturdily muscled hindquarters, essential for the fast departure required in short races. It serves as a polo pony equally well as for ranch work.

The Morgan horse originated from a stallion given to Justin Morgan of Vermont around 1795. This breed has become a most versatile horse for riding, pulling carriages, farm labour, and cattle cutting. It was the ideal army charger. It stands about 15 hands (152.4 cm, or 60 inches) high and is robust, good-natured, willing, and intelligent. Its coat is dark brown or liver chestnut.

Appaloosa is a colour breed said to have descended in the Nez Percé Indian territory of North America from wild mustangs, which in turn descended from Spanish horses brought to the New World by explorers. The Appaloosa is 14.2 to 15.2 hands (about 147 to 157.5 cm, or 58 to 62 inches) high, of sturdy build, and of most diverse use; it is especially good in farmwork. There are various breeds of spotted horses in Europe and Asia, and the actual source of the spotting pattern in the Appaloosa is uncertain.

American breeders have also developed several horses that have specialized gaits. These gaited breeds include the American Saddlebred horse, the Tennessee Walking Horse, and the Missouri Fox Trotting Horse.

The American Saddlebred horse has a small head and spectacular high-stepping movements. It is trained for either three or five gaits. The three-gaited horses perform the walk, trot, and canter; the five-gaited horses in addition perform the rack, a quick, high-stepping four-beat gait, and the slow gait, a somewhat slower form of the rack. Since these horses are used mainly for shows, their hooves are kept rather long, and the muscles of the tail are often clipped so that the base of the tail is carried high. Chestnut and bay are the usual colours.

The Tennessee Walking Horse—a breed derived partially from the Thoroughbred, Standardbred, Morgan, and American Saddlebred horse—serves as a comfortable riding mount used to cover great distances at considerable speed. Its specialty is the running walk, a long and swift stride. Bay is the most common colour.

The Missouri Fox Trotting Horse, a breed developed to cover the rough terrain of the Ozark region, is characterized by an unusual gait, called the fox-trot, in which the front legs move at a walk while the hind legs perform a trot. The most common colours for this breed are sorrel and chestnut sorrel.

## Other Light Breeds

The English Hackney is a light carriage horse, influenced by the Thoroughbred and capable of covering distances of 12 to 15 miles (19 to 24 km) per hour at the trot and canter. It measures 15.2 to 15.3 hands (about 157.5 to 160 cm, or 62 to 63 inches) high and is appreciated for its high knee action.

The Cleveland Bay carriage horse, up to 17 hands (about 172.7 cm, or 68 inches) high and generally bay in colour, is similar to the Yorkshire Coach horse. Both breeds are now used for the equestrian event of carriage driving.

Other versatile breeds include the German Holstein, Hanoverian, and East Prussian (Trakehner), which serve equally well for riding, light labour, and carriage. These horses, 16 to 18 hands (about 162.6 to 182.9 cm, or about 64 to 72 inches) high and of all colours, are now mostly bred for sport.

Hanoverian stallion with dark bay coat.

The Andalusian, a high-stepping spirited horse, and the small but enduring Barb produced the Lipizzaner, which was named after the stud farm founded near Trieste, Italy, in 1580. Originally of all colours, the Lipizzaner is gray or, now exceptionally, bay. It is small, rarely over 15 hands (152.4 cm, or 60 inches) high, and of powerful build but with slender legs and a long silky mane and tail. Intelligence and sweetness of disposition as well as gracefulness destined it for academic horsemanship, notably as practiced at the Spanish Riding School of Vienna.

Lipizzaner horse.

In figure, a Lipizzaner, or Lipizzan, horse performing a capriole, "the leap of the goat," in which the horse jumps into the air from a raised starting position. The breed was named after the stud farm at Lippiza, which was founded near Trieste, Italy, in 1580.

Andalusian stallion with dark gray coat.

## Heavy Breeds

The horses used for heavy loads and farm labour descended from the ancient war horses of the Middle Ages. These breeds—including the English Shire (the world's largest horse), Suffolk, and Clydesdale; the French Percheron; the Belgian horse; the German Noriker; and the Austrian Pinzgauer—are now little used for their original purpose, having been almost entirely replaced by the tractor. They usually measure well over 16 hands (about 162.6 cm, or 64 inches) high, some more than 19 hands (about 193 cm, or 76 inches). They are of all colours, sometimes spotted, and generally have a very calm temperament. Many of these breeds are rare and endangered at present.

## Ponies

Ponies are any horses other than Arabians that are shorter than 14.2 hands (about 147 cm, or 58 inches). They are generally very sturdy, intelligent, energetic, and sometimes stubborn. The coat is of all colours, mainly dark, and the mane and tail are full. Ponies are used for pulling carriages and pack loads and as children's riding horses or pets. There are numerous varieties, including the Welsh, Dartmoor, Exmoor, Connemara, New Forest, Highland, Dale, Fell, Pony of the Americas, Shetland (under 7 hands [71.1 cm, or 28 inches] high), Iceland, and Norwegian. Ponies of the warmer countries include the Indian, Java, Manila, and Argentina.

An Icelandic horse moving swiftly at the tölt, a smooth four-beat, lateral running walk.

Originating in the South Tyrol, the Haflinger is a mountain pony, enduring, robust, and versatile, used for all farm labour, for pulling a carriage or sledge, and for pack hauling. It is chestnut with a flaxen mane and tail.

A team of Haflingers at a driving demonstration.

Some breeds of ponies, such as the Caspian, are short but have the body proportions of a horse instead of the shorter legs relative to body size of the true ponies of northern Europe.

# References

- Woods, Katie (30 July 2015). "How to determine if cattle are bulls, steers, cows or heifers - Farm and Dairy". Farm and Dairy. Retrieved December 27, 2017

- Cattle-livestock, animal: britannica.com, Retrieved 24 January, 2019

- Booth, K. K.; Katz, L. S. (2000). "Role of the vomeronasal organ in neonatal offspring recognitions in sheep". Biol. Reprod. 63 (3): 353–358. Doi:10.1095/biolreprod63.3.953

- Basnet, S.; Schneider, M.; Gazit, A.; Mander, G.; Doctor, A. (2010). "Fresh Goat's Milk for Infants: Myths and Realities—A Review". Pediatrics. 125 (4): e973–e977. Doi:10.1542/peds.2009-1906. PMID 20231186. Retrieved 14 July 2010

- Merriam-Webster Unabridged (MWU). (Online subscription-based reference service of Merriam-Webster, based on Webster's Third New International Dictionary, Unabridged. Merriam-Webster, 2002.) Headword donkey. Retrieved September 2007

- Horse, animal: britannica.com, Retrieved 25 February 2019 Prinsen, M. K. (2006). "The Draize Eye Test and in vitroalternatives: A left-handed marriage?". Toxicology in Vitro. 20(1): 78–81. Doi:10.1016/j.tiv.2005.06.030. PMID 16055303

# 3
# Animal Husbandry

The branch of agriculture which is concerned with animals that are raised for milk, eggs, fiber and meat is referred to as animal husbandry. The important aspects of animal husbandry include animal breeding and animal feed. The chapter closely examines these key aspects of animal husbandry to provide an extensive understanding of the subject.

Animal husbandry is the Controlled cultivation, management, and production of domestic animals, including improvement of the qualities considered desirable by humans by means of breeding. Animals are bred and raised for utility (e.g., food, fur), sport, pleasure, and research.

Animal husbandry began in the so-called Neolithic ('new stone') Revolution around 10,000 years ago but may have begun much earlier. It has been speculated that human beings used fire to cook food 1.5 million years ago but the only archaeological evidence obtained thus far sets the date of the use of fire for cooking at 12,500 years ago as indicated by the discovery of clay cooking pots in East Asia and Mesopotamia.

Shortly after this date, evidence of domesticated animal bones left over from human social gatherings such as dinners emerges; said bones having been discovered in excavations of fire pits in ancient kitchens. Though domestication of animals was probably common earlier, it is certain that goats and sheep were domesticated throughout Asia by 8000 BCE. Wheat was domesticated and in wide use in Mesopotamia by 7700 BCE, goats by 7000 BCE, sheep by 6700 BCE, and pigs by 6500 BCE. By the time of the settlement of the first Mesopotamian city of Eridu in 5400 BCE, animal husbandry was widely practiced and domesticated animals used in the work force (such as in ploughing) as pets, and as a food source. Horses were tamed by 4000 BCE and, in time, became an important component in warfare in drawing the great chariots of the various nation-states. Eventually, elephants, tigers, and lions were employed on the battlefield; particularly in the latter cases of the Persian campaigns, the Indian resistance to Alexander the Great, and, most famously, by Hannibal of Carthage against the Romans.

## Animal Breeding

Animal breeding is the selective mating of animals to increase the possibility of obtaining desired traits in their offspring. It has been performed with most domesticated animals, especially cats and dogs, but its main use has been to breed better agricultural stock. More modern techniques involve a wide variety of laboratory methods, including the modification of embryos, sex selection, and genetic engineering.

These procedures are beginning to supplant traditional breeding methods, which focus on selectively combining and isolating livestock strains. In general, the most effective strategy for isolating traits is by selective inbreeding; but different strains are sometimes crossed to take advantage of hybrid vigor and to forestall the negative results of inbreeding, which include reduced fertility, low immunity, and the development of genetic abnormalities.

## The Genetic basis of Animal Breeding

Breeders engage in genetic "experiments" each time they plan a mating. The type of mating selected depends on the goals. To some breeders, determining which traits will appear in the offspring of a mating is like rolling the dice—a combination of luck and chance. For others, producing certain traits involves more skill than luck—the result of careful study and planning. Dog breeders, for example, have to understand how to manipulate genes within their breeding stock to produce the kinds of dogs they want. They have to first understand dogs as a species, then dogs as genetic individuals.

Once the optimal environment for raising an animal to maturity has been established (i.e., the proper nutrition and care has been determined) the only way to manipulate an animal's potential is to manipulate its genetic information. In general, the genetic information of animals is both diverse and uniform: diverse in the sense that a population will contain many different forms of the same gene (for instance, the human population has 300 different forms of the protein hemoglobin); and uniform in the sense that there is a basic physical expression of the genetic information that makes, for instance, most goats look similar to each other.

In order to properly understand the basis of animal breeding, it is important to distinguish between genotype and phenotype. Genotype refers to the information contained in an animal's DNA, or genetic material. An animal's phenotype is the physical expression of its genotype. Although every creature is born with a fixed genotype, the phenotype is a variable influenced by many factors in the animal's environment and development. For example, two cows with identical genotypes could develop quite different phenotypes if raised in different environments and fed different foods.

The close association of environment with the expression of the genetic information makes animal breeding a challenging endeavor, because the physical traits a breeder desires to selectively breed for cannot always be attributed entirely to the animal's genes. Moreover, most traits are due not just to one or two genes, but to the complex interplay of many different genes.

DNA consists of a set of chromosomes; the number of chromosomes varies between species (humans, for example, have 46). Mammals (and indeed most creatures) have two copies of each chromosome in the DNA (this is called diploidy). This means there are two copies of the same gene in an animal's DNA. Sometimes each of these will be partially expressed. For example, in a person having one copy of a gene that codes for normal hemoglobin and one coding for sickle-cell hemoglobin, about half of the hemoglobin will be normal and the other half will be sickle-cell. In other cases, only one of the genes can be expressed in the animal's phenotype. The gene expressed is called dominant, and the gene that is not expressed is called recessive. For instance, a human being could have two copies of the gene coding for eye color; one of them could code for blue, one for brown. The gene coding for brown eyes would be dominant, and the individual's eyes would be brown. But the blue-eyes gene would still exist, and could be passed on to the person's children.

Most of the traits an animal breeder might wish to select will be recessive, for the obvious reason that if the gene were always expressed in the animals, there would be no need to breed for it. If a gene is completely recessive, the animal will need to have two copies of the same gene for it to be expressed (in other words, the animal is homozygous for that particular gene). For this reason, animal breeding is usually most successful when animals are selectively inbred. If a bull has two copies of a gene for a desirable recessive trait, it will pass one copy of this gene to each of its offspring. The other copy of the gene will come from the cow, and assuming it will be normal, none of the offspring will show the desirable trait in their phenotype. However, each of the offspring will have a copy of the recessive gene. If they are then bred with each other, some of their offspring will have two copies of the recessive gene. If two animals with two copies of the recessive gene are bred with each other, all of their offspring will have the desired trait.

There are disadvantages to this method, although it is extremely effective. One of these is that for animal breeding to be performed productively, a number of animals must be involved in the process. Another problem is that undesirable traits can also mistakenly be selected for. For this reason, too much inbreeding will produce sickly or unproductive stock, and at times it is useful to breed two entirely different strains with each other. The resulting offspring are usually extremely healthy; this is referred to as "hybrid vigor." Usually hybrid vigor is only expressed for a generation or two, but crossbreeding is still a very effective means to combat some of the disadvantages of inbreeding.

Another practical disadvantage to selective inbreeding is that the DNA of the parents is altered during the production of eggs and sperm. In order to make eggs and sperm, which are called gametes, a special kind of cell division occurs called meiosis, in which cells divide so that each one has half the normal number of chromosomes (in humans, each sperm and egg contains 23 chromosomes). Before this division occurs, the two pairs of chromosomes wrap around each other, and a phenomenon known as crossing over takes place in which sections of one chromosome will be exchanged with sections of the other chromosome so that new combinations are generated.

The problem with crossing over is that some unexpected results can occur. For instance, the offspring of a bull homozygous for two recessive but desirable traits and a cow with "normal" genes will all have one copy of each recessive gene. But when these offspring produce gametes, one recessive gene may migrate to a different chromosome, so that the two traits no longer appear in one gamete. Since most genes work in complicity with others to produce a certain trait, this can make the process of animal breeding very slow, and it requires many generations before the desired traits are obtained—if ever.

## Economic Considerations

There are many reasons why animal breeding is of paramount importance to those who use animals for their livelihood. Cats have been bred largely for aesthetic beauty; many people are willing to pay a great deal of money for a Siamese or Persian cat, even though the affection felt for a pet has little to do with physical appearance. But the most extensive animal breeding has occurred in those areas where animals have been used to serve specific practical purposes. For instance, most dog breeds are the result of a deliberate attempt to isolate traits that would produce better hunting and herding dogs (although some, like toy poodles, were bred for traits that would make them desirable pets). Horses have also been extensively bred for certain useful qualities; some for size and strength, some for speed. But farm animals, particularly food animals, have been the subject of the most intensive breeding efforts.

The physical qualities of economic importance in farm animals vary for each species, but a generalized goal is to eliminate the effects of environment and nutrition. An ideal strain of milk cow, for instance, would produce a large amount of high-quality milk despite the type of food it is fed and the environment in which it is reared. Thus, animals are generally all bred for feed efficiency, growth rate, and resistance to disease. However, a pig might be bred for lean content in its meat, while a hen would be bred for its laying potential. Many cows have been bred to be hornless, so they cannot inadvertently or deliberately gore each other.

Although maximum food production is always a major goal, modern animal breeders are also concerned about nutritional value and the ability of animals to survive in extreme environments. Many parts of the world are sparsely vegetated or have harsh climatic conditions, and a high-efficiency producer able to endure these environments would be extremely useful to the people who live there. In addition, many people of industrialized countries are concerned not about food availability but about the quality of this food; so breeders seek to eliminate the qualities that make meat or milk or eggs or other animal products unhealthy, while enhancing those qualities that make them nutritious.

## Modern Methods in Biotechnology

Although earlier animal breeders had to confine themselves to choosing which of their animals should mate, modern technological advances have altered the face of animal breeding, making it both more selective and more effective. Techniques like genetic engineering, embryo manipulation, artificial insemination, and cloning are becoming more refined. Some, like artificial insemination and the manipulation of embryos to produce twins, are now used habitually. Others, such as genetic engineering and cloning, are the subject of intense research and will probably have a great impact on future animal breeding programs.

## Artificial Insemination

Artificial insemination is the artificial introduction of semen from a male with desirable traits into females of the species to produce pregnancy. This is useful because a far larger number of offspring can be produced than would be possible if the animals were traditionally bred. Because of this, the value of the male as breeding stock can be determined much more rapidly, and the use of many different females will permit a more accurate evaluation of the heredit-ability of the desirable traits. In addition, if the traits produced in the offspring do prove to be advantageous, it is easier to disperse them within an animal population in this fashion, as there is a larger breeding stock available. One reason artificial insemination has been an extremely important tool is that it allowed new strains of superior stock to be introduced into a supply of animals in an economically feasible fashion.

The process of artificial insemination requires several steps. Semen must be obtained and effectively diluted, so that the largest number of females can be inseminated (consistent with a high probability of pregnancy). The semen must be properly stored so that it remains viable. The females must be tested before the sample is introduced to ensure they are fertile, and, following the procedure, they must be tested for pregnancy to determine its success. All these factors make artificial insemination more expensive and more difficult than traditional breeding methods, but the processes have been improved and refined so that the economic advantages far outweigh the procedural disadvantages. Artificial insemination is the most widely applied breeding technique.

## Embryo Manipulation

To understand the techniques of embryo manipulation, it is important to understand the early stages of reproduction. When the egg and sperm unite to form a zygote, each of the parents supply half of the chromosomes necessary for a full set. The zygote, which is a single cell, then begins to reproduce itself by the cellular division process called mitosis, in which each chromosome is duplicated before separation so that each new cell has a full set of chromosomes. This is called the morula stage, and the new cells are called blastomeres. When enough cells have been produced (the number varies from species to species), cell differentiation begins to take place. The first differentiation appears to be when the blastocyst is formed, which is an almost hollow sphere with a cluster of cells inside; and the differentiation appears to be between the cells inside, which become the fetus, and the cells outside, which become the fetal membranes and placenta. However, the process is not yet entirely understood, and there is some variation between species, so it is difficult to pinpoint the onset of differentiation, which some scientists believe occurs during blastomere division.

During the first stages of cell division, it is possible to separate the blastomeres with the result that each one develops into a separate embryo. Blastomeres with this capability are called totipotent. The purpose of this ability of a single blastomere to produce an entire embryo is probably to safeguard the process of embryo development against the destruction of any of the blastomeres. In theory, it should be possible to produce an entire embryo from each blastomere (and blastomeres are generally totipotent from the four to eight cell stage), but in practice it is usually only possible to produce two embryos. That is why this procedure is generally referred to as embryo splitting rather than cloning, although both terms refer to the same thing (cloning is the production of genetically identical embryos, which is a direct result of embryo splitting). Interestingly, although the embryos produced from separated blastomeres usually have fewer cells than a normal embryo, the resulting offspring fall within the normal range of size for the species.

It is also possible to divide an embryo at other stages of development. For instance, the time at which embryo division is most successful is after the blastocyst has formed. Great care must be taken when dividing a blastocyst, since differentiation has already occurred to some extent, and it is necessary to halve the blastocyst very precisely.

Another interesting embryonic manipulation is the creation of chimeras. These are formed by uniting two different gametes, so that the embryo has two distinct cell lineages. Chimeras do not combine the genetic information of both lineages in each cell. Instead, they are a patchwork of cells containing one lineage or the other. For this reason, the offspring of chimeras are from one distinct genotype or the other, but not from both. Thus chimeras are not useful for creating new animal populations beyond the first generation. However, they are extremely useful in other contexts. For instance, while embryo division as described above is limited in the number of viable embryos that can be produced, chimeras can be used to increase the number. After the blastomeres are separated, they can be combined with blastomeres of a different genetic lineage. The additional tissue, in fact, increases the new embryos' survival rate. For some reason only a small percentage of the resulting embryos are chimeric; this is thought to be because only one cell lineage develops into the cells inside the blastocyst, while the other lineage forms extra-embryonic tissue. Scientists believe that the more advanced cells are more likely to form the inner cells.

Chimeras could also be used to breed endangered species. Because of different uterine biochemical environments and the different regulatory mechanisms for fetal development, only very closely related species are able to bear each other's embryos to term. For example, when a goat is implanted with a sheep embryo or the other way around, the embryo is unable to develop properly. This problem can perhaps be surmounted by creating chimeras in which the placenta stems from the cell lineage of the host species. The immune system of an animal attacks tissue it recognizes as foreign, but it is possible that the mature chimeras would be compatible with both the host and the target species, so that it could bear either embryo to term. This has already proven true in studies with mice.

A further technique being developed to manipulate embryos involves the creation of uniparental embryos and same-sex mating. In the former case, the cell from a single gamete is made to undergo mitosis, so that the resulting cell is completely homo-zygous. In the latter case, the DNA from two females (parthogenesis) or two males (androgenesis) is combined to form cells that have only female- or male-derived DNA. These zygotes cannot be developed into live animals, as genetic information from male- and female-derived DNA is necessary for embryonic development.

## Genetic Engineering

Genetic engineering can produce transgenic animals—those that have had a new gene inserted directly into their DNA. The procedure involves microinjection of the desired gene into the nucleus of fertilized eggs. Although success rates vary, in many cases the new gene is reproduced in all developing cells and can be transcribed, or read, and utilized by the cell. This is a startling breakthrough in animal breeding, because it means a specific trait can be incorporated into a population in a single generation, rather than the several generations this takes with conventional breeding techniques.

There are some serious limitations to the procedure, however. The first is that many genes must work together to produce the very few traits a breeder would like to include in an animal population. Although it might some day be possible to incorporate any number of genes into an embryo's DNA, the complex interplay of genes is not understood very well, and the process of identifying all of those related to a desired trait is costly and time-consuming.

Another problem in the production of transgenic animals is that they pass their modified DNA on to their offspring with varying success rates and unpredictable results. In some cases, the new gene is present in the offspring but is not utilized. Or it may be altered or rearranged in some way, probably during the process of gamete production.

These factors make it difficult to produce a strain of transgenic animals. However, with further research into the mechanism by which the gene is incorporated into the genome, and by mapping the target animal genome and identifying the genes responsible for various traits, genetic engineering will no doubt become a major tool for improving animal strains.

## Sex Selection

It would be extremely useful if a breeder were able to predetermine the sex of each embryo produced, because in many cases one sex is preferred. For instance, in a herd of dairy cows or a flock of laying hens, females are the only commercially useful sex. When the owner of a dairy herd has inseminated a cow at some expense, this issue becomes more crucial. In some cases, an animal is being bred specifically for use as breeding stock; in this case, it is far more useful to produce a male that can be bred with multiple females than a female, which can only produce a limited number of offspring.

Whether or not an animal is male or female is determined by its sex chromosomes, which are called X and Y chromosomes. An animal with two X chromosomes will develop into a female, while an animal with one X and one Y chromosome will become a male. In mammals, the sex of the offspring is almost always determined by the male parent, because the female can only donate an X chromosome, and it is the presence or absence of the Y chromosome that causes maleness (this is not true in, for instance, birds; in that case the female has two different sex chromosomes). Using cell sorters, researchers are now able to separate spermatozoa into X and Y components. Although such "sex-sorted" sperm produces fewer pregnancies than nonsorted sperm, more than 90% of the offspring produced are of the intended sex.

## Breeding Stock

Breeding stock is a group of animals used for the purpose of planned breeding. When individuals are looking to breed animals, they look for certain valuable traits in purebred animals, or may intend to use some type of crossbreeding to produce a new type of stock with different, and presumably super abilities in a given area of endeavor. For example, when breeding swine for meat, the "breeding stock should be sound, fast growing, muscular, lean, and reproductively efficient." The "subjective selection of breeding stock" in horses has led to many horse breeds with particular performance traits.

## Selection of Breeding Stock

Selection is used as a tool for livestock improvement. A breeding stock is a group of males and females which act as parents of future generations.

Selection is the process of allowing certain animals to be parents of future generations while culling others.

Culling is the removal of animals which do not perform to the desired level, from the herd. The animals retained have certain desirable characteristics which make them produce more.

Selected animals make up the breeding stock.

The breeding stock should pass the good qualities to their offsprings for better performance, to improve the livestock.

Selection process repeated for many generations increases chances of formation of desirable qualities in an animal. Genetically termed as gene frequency(occurrence of the genes that carry desirable characteristics.) Selection increases occurrence of desirable genes and decreases occurance of undesirable genes. During selection, the characteristics to be selected for are first studied closely to ascertain that it is not influenced by the environment, but mainly by the genetic make-up.

Selection helps improve characteristics which are highly heritable.

Heritability means the likelihood of a particular trait to be transmitted to the offspring and they are strongly inherited. A character like milk yield is lowly heritable, i.e. it is weakly inherited and a bigger percentage of the character is affected by the environment.

The degree to which selection affects a character depends on the following factors: The heritability of the character, The intensity with which the selection is done and the interval between generations and kind of selection being practiced.

## Factors to Consider when Selecting a Breeding Stock

- Age
- Level of performance
- Physical Fitness
- Health
- Body Conformation
- Temperament or Behavior
- Quality of products
- Mothering Ability
- Adaptability
- Prolificacy

## Age

- Young animals.
- Those that have not parturated for more than 3-times, should be selected.
- They have a longer productive life.
- Old animals are poor breeders and low producers.
- Production and breeding efficiency decline with age.

## Level of Performance

- Animals with highest production level selected.
- Performance best indicated by records.

Good Performance of animal indicated by:

- High milk, wool and egg production.
- Good mothering ability.
- High prepotency which is the ability of a parent to pass good qualities to their offsprings.
- The animals with poor performance should be culled.
- Good records kept and used by the farmer for this purpose.

## Physical Fitness

Animals selected should be free from any physical defect:

- Mono-eyed,
- Limping,
- Irregular number of teats,
- Scrotal hernia,
- Defective and weak backline.

## Health

- Sick animals do not breed well and are expensive to keep.
- Animals that are resistant to diseases pass these characteristics to their offsprings.

## Body Conformation

- Animals for breeding to be selected according to proper body conformation.
- A dairy cow should be wedge-shaped with a large udder, thin legs, long neck.

## Temperament or Behavior

- Animals with bad behaviors should be culled. e.g Cannibalism, egg eating, aggressiveness, kicking.

## Quality of Products

- Select animals that give products of high quality such as meat, wool, eggs, milk.

## Mothering Ability

- Animals selected should have a good mothering ability.

- That is animals with good natural instinct towards their young ones.

- This will enable them to rear the young ones up to weaning.

## Adaptability

Animals selected should be well adapted to the prevailing climatic condition in the area e.g Ardi and semi arid areas.

## Prolificacy

- Animals selected should be highly prolific.

- That is, animals with the ability to give birth to many offsprings at a time(larger litter).

- This is a quality that should be considered when selecting pigs and rabbits.

- The ancestry records assist to choose the prolific breedsfor mating

## Selection of Cattle and Sheep

## Selection in Cattle

Consider the following:

- Level of performance which include:

  ○ Milk yield buter content.

  ○ Length of lactation period.

  ○ Calving intervals.

Age of the animal, fertility, physical fitness, health of the animal, body conformation and suitability of the enterprise-milk or beef.

## Selection in Sheep

Consider the following:

- Level of performance which includes:

  ○ Mothering ability

  ○ Growth rate

  ○ Wool quality

  ○ Carcass quality

- Twining rate Age
- Suitability to the enterprise-wool or mutton
- Flocking instinct Health of the animal
- Physical fitness
- Inheritable defects
- Fertility

## Selection in Goats

Consider the following:

- Fertility
- Mothering ability
- Growth rate
- Twining rate
- Carcass quality/dressing percentage
- Growth rate
- Suitability to the enterprise - milk or mutton
- Health of the animal
- Age

## Selection in Pigs

Consider the following:

- Carcass quality/dressing percentage
- Suitability to the enterprise (bacon or pork)
- Growth rate
- Health of the animal
- Mothering ability
- Prolificacy
- Number of teats
- Temperament

- Body formation

- Age

- Heredity defects

## Selection in Camels

- Health of the animal

- Age

- Temperament

- Foraging ability

- Fertility

- Level of performance-milk, meat, fur and transport

## Methods of Selection

- Mass selection - Animals with superior characteristics (highly heritable breeds) are selected from a herd and then allowed to mate among each other at random. The offsprings will show higher performance than their parents. This is because mass selection increases the occurrence of the desirable genes in a population.

- Progeny testing - Is a offspring resulting from selected parents ( Family selection).In this method a group of progenies (offrprings) are used as an aid to increase accuracy in the selection ofa breeding stock. This is method is used when the character to be selected is of low heritability and expressedby one sex only.

This method takes upto nine years for the results to be seen.

## Contemporary Comparison

Contemporaries refers to other heifers in the herd sired by the same bull. This is a progey tasting method which involves comparison of average production of daughters (Heifers) of each bull with that of the other heifer refered to as contemporaries. In this methods it is assumed that thedifference between the herds of the same breed are non-genetic in origin.

## Advantages

- It is possible to compare heifers of different ages in different locations worldwide.

- It eliminates difference brought about by the environment.

- it is possible to make direct comparison of stut bulls at different artificial insemiation centres.

- It is accurate since we are using a large herd of animals.

## Purebred

Purebreds, also called purebreeds, are cultivated varieties or *cultivars* of an animal species, achieved through the process of selective breeding. When the lineage of a purebred animal is recorded, that animal is said to be *pedigreed*.

The term *purebred* is occasionally confused with the proper noun *Thoroughbred*, which refers exclusively to a specific breed of horse, one of the first breeds for which a written national stud book was created since the 18th century. Thus a purebred animal should never be called a "thoroughbred" unless the animal actually is a registered Thoroughbred horse.

## True Breeding

In the world of selective animal breeding, to "breed true" means that specimens of an animal breed will breed true-to-type when mated like-to-like; that is, that the progeny of any two individuals in the same breed will show consistent, replicable and predictable characteristics. A puppy from two purebred dogs of the same breed, for example, will exhibit the traits of its parents, and not the traits of all breeds in the subject breed's ancestry.

However, breeding from too small a gene pool, especially direct inbreeding, can lead to the passing on of undesirable characteristics or even a collapse of a breed population due to inbreeding depression. Therefore, there is a question, and often heated controversy, as to when or if a breed may need to allow "outside" stock in for the purpose of improving the overall health and vigor of the breed.

Because pure-breeding creates a limited gene pool, purebred animal breeds are also susceptible to a wide range of congenital health problems. This problem is especially prevalent in competitive dog breeding and dog show circles due to the singular emphasis on aesthetics rather than health or function. Such problems also occur within certain segments of the horse industry for similar reasons. The problem is further compounded when breeders practice inbreeding. The opposite effect to that of the restricted gene pool caused by pure-breeding is known as hybrid vigor, which generally results in healthier animals.

## Pedigrees

A purebred Arabian horse.

A pedigreed animal is one that has its ancestry recorded. Often this is tracked by a major registry. The number of generations required varies from breed to breed, but all pedigreed animals have papers from the registering body that attest to their ancestry.

The word "pedigree" appeared in the English language in 1410 as "pee de Grewe", "pedegrewe" or "pedegru", each of those words being borrowed to the Middle French "pié de grue", meaning "crane foot". This comes from a visual analogy between the trace of the bird's foot and the three lines used in the English official registers to show the ramifications of a genealogical tree.

Sometimes the word *purebred* is used synonymously with *pedigreed*, but purebred refers to the animal having a known ancestry, and pedigree refers to the written record of breeding. Not all purebred animals have their lineage in written form. For example, until the 20th century, the Bedouin people of the Arabian peninsula only recorded the ancestry of their Arabian horses via an oral tradition, supported by the swearing of religiously based oaths as to the asil or "pure" breeding of the animal. Conversely, some animals may have a recorded pedigree or even a registry, but not be considered "purebred". Today the modern Anglo-Arabian horse, a cross of Thoroughbred and Arabian bloodlines, is considered such a case.

## Purebreds by Animal

### Purebred Horses

The domestication of the horse resulted in a small number of domesticated stallions (possibly a single male ancestor) being crossed on wild mares that had adapted to local conditions. This ultimately produced horses of four basic body types, once thought to be wild prototypes, but now considered to be landraces. Many of these animals were then bred true to original type by selected breeding, though emphasizing certain inherent traits (such as a good temperament, suitable to training by humans) to a greater degree than others. In other cases, horses of different body types were cross bred until a desired characteristic was achieved and bred true.

Written and oral histories of various animals or pedigrees of certain types of horse have been kept throughout history, though breed registry stud books trace only to about the 13th century, at least in Europe, when pedigrees were tracked in writing, and the practice of declaring a type of horse to be a breed or a purebred became more widespread.

Certain horse breeds, such as the Andalusian horse and the Arabian horse, are claimed by aficionados of the respective breeds to be ancient, near-pure descendants from an ancient wild prototype, though mapping of the horse genome as well as the mtDNA and y-DNA of various breeds has largely disproved such claims.

### Purebred Livestock

Most domesticated farm animals also have true-breeding breeds and breed registries, particularly cattle, sheep, goats, rabbits, and pigs. While animals bred strictly for market sale are not always purebreds, or if purebred may not be registered, most livestock producers value the presence of purebred genetic stock for the consistency of traits such animals provide. It is common for a farm's male breeding stock in particular to be of purebred, pedigreed lines.

Purebred Barbados Blackbelly hair sheep.

In cattle, some breeders associations make a difference between "purebred" and "full blood". Full blood cattle are fully pedigreed animals, where every ancestor is registered in the herdbook and shows the typical characteristics of the breed. Purebred are those animals that have been bred-up to purebred status as a result of using full blood animals to cross with an animal of another breed. The breeders association rules the percentage of fullblood genetics required for an animal to be considered purebred, usually above 87.5%.

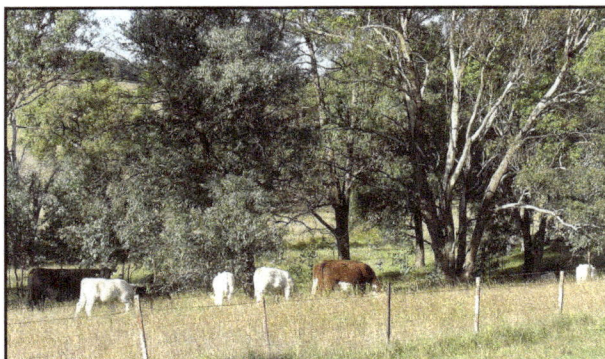
Charolais calves that were transferred as embryos, with their Angus and Hereford recipient mothers.

Artificial breeding via artificial insemination or embryo transfer is often used in sheep and cattle breeding to quickly expand, or improve purebred herds. Embryo transfer techniques allow top quality female livestock to have a greater influence on the genetic advancement of a herd or flock in much the same way that artificial insemination has allowed greater use of superior sires.

## Wild Species, Landraces and Purebred Species

Breeders of purebred domesticated species discourage crossbreeding with wild species, unless a deliberate decision is made to incorporate a trait of a wild ancestor back into a given breed or strain. Wild populations of animals and plants have evolved naturally over millions of years through a process of natural selection in contrast to human controlled Selective breeding or artificial selection for desirable traits from the human point of view. Normally, these two methods of reproduction operate independently of one another. However, an intermediate form of selective breeding, wherein animals or plants are bred by humans, but with an eye to adaptation to natural region-specific conditions and an acceptance of natural selection to weed out undesirable traits, created many ancient domesticated breeds or types now known as landraces.

Many times, domesticated species live in or near areas which also still hold naturally evolved, region-specific wild ancestor species and subspecies. In some cases, a domesticated species of plant or animal may become feral, living wild. Other times, a wild species will come into an area inhabited by a domesticated species. Some of these situations lead to the creation of hybridized plants or animals, a cross between the native species and a domesticated one. This type of crossbreeding, termed genetic pollution by those who are concerned about preserving the genetic base of the wild species, has become a major concern. Hybridization is also a concern to the breeders of purebred species as well, particularly if the gene pool is small and if such crossbreeding or hybridization threatens the genetic base of the domesticated purebred population.

The concern with genetic pollution of a wild population is that hybridized animals and plants may not be as genetically strong as naturally evolved region specific wild ancestors wildlife which can survive without human husbandry and have high immunity to natural diseases. The concern of purebred breeders with wildlife hybridizing a domesticated species is that it can coarsen or degrade the specific qualities of a breed developed for a specific purpose, sometimes over many generations. Thus, both purebred breeders and wildlife biologists share a common interest in preventing accidental hybridization.

## Crossbreed

A crossbreed is an organism with purebred parents of two different breeds, varieties, or populations. Crossbreeding, sometimes called "designer crossbreeding", is the process of breeding such an organism, often with the intention to create offspring that share the traits of both parent lineages, or producing an organism with hybrid vigor. While crossbreeding is used to maintain health and viability of organisms, irresponsible crossbreeding can also produce organisms of inferior quality or dilute a purebred gene pool to the point of extinction of a given breed of organism.

A domestic animal of unknown ancestry, where the breed status of only one parent or grandparent is known, may also be called a crossbreed though the term "mixed breed" is technically more accurate. Outcrossing is a type of crossbreeding used within a purebred breed to increase the genetic diversity within the breed, particularly when there is a need to avoid inbreeding.

A Zebroid, a crossbreed between a zebra and a horse.

In animal breeding, *crossbreeds* are crosses within a single species, while *hybrids* are crosses between different species. In plant breeding terminology, the term *crossbreed* is uncommon, and no universal term is used to distinguish hybridization or crossing within a population from those between populations, or even those between species.

## Crossbreeds in Specific Animals

### Cattle

In cattle, there are systems of crossbreeding. In many crossbreeds, one animal is larger than the other. One is used when the purebred females are particularly adapted to a specific environment, and are crossed with purebred bulls from another environment to produce a generation having traits of both parents.

### Sheep

The large number of breeds of sheep, which vary greatly, creates an opportunity for crossbreeding to be used to tailor production of lambs to the goal of the individual stockman.

### Llamas

Results of crossbreeding classic and woolly breeds of llama are unpredictable. The resulting offspring displays physical characteristics of either parent, or a mix of characteristics from both, periodically producing a fleeced llama. The results are increasingly unpredictable when both parents are crossbreeds, with possibility of the offspring displaying characteristics of a grandparent, not obvious in either parent.

### Horses

The National Show Horse was developed from crossbreeding programs in the 1970s and 1980s that blended Arabian horse and American Saddlebred bloodlines.

Crossbreeding in horses is often done with the intent of ultimately creating a new breed of horse. One type of modern crossbreeding in horses is used to create many of the warmblood breeds. Warmbloods are a type of horse used in the sport horse disciplines, usually registered in an open stud book by a studbook selection procedure that evaluates conformation, pedigree and, in some animals, a training or performance standard. Most warmblood breeds began as a cross of draft horse breeds on Thoroughbreds, but have, in some cases, developed over the past century to the point where they are considered to be a true-breeding population and have a closed stud book. Other types of recognized crossbreeding include that within the American Quarter Horse, which will register horses with one Thoroughbred parent and one registered Quarter Horse parent in the "Appendix" registry, and allow such animals full breed registration status as Quarter Horses if they meet a certain performance standard. Another well-known crossbred horse is the Anglo-Arabian, which may be produced by a purebred Arabian horse crossed on a Thoroughbred, or by various crosses of Anglo-Arabians with other Anglo-Arabians, as long as the ensuing animal never has more than 75% or less than 25% of each breed represented in its pedigree.

## Hybrid Animals

A hybrid animal is one with parentage of two separate species, differentiating it from crossbred animals, which have parentage of the same species. Hybrids are usually, but not always, sterile.

One of the most ancient types of hybrid animal is the mule, a cross between a female horse and a male donkey. The liger is a hybrid cross between a male lion and female tiger. The yattle is a cross between a cow and a yak. Other crosses include the tigon (between a male tiger and female lion) and yakalo (between a yak and buffalo). The Incas recognized that hybrids of *Lama glama* (llama) and *Vicugna pacos* (alpaca) resulted in a hybrid with none of the advantages of either parent.

At one time it was thought that dogs and wolves were separate species, and the crosses between dogs and wolves were called wolf hybrids. Today wolves and dogs are both recognized as *Canis lupus*, but the old term "wolf hybrid" is still used.

## Mixed Breeds

A mixed-breed animal is defined as having undocumented or unknown parentage, while a cross-breed generally has known, usually purebred parents of two distinct breeds or varieties. A dog of unknown parentage is often called a mixed-breed dog, "mutt" or "mongrel." A cat of unknown parentage is often referred to as domestic short-haired or domestic long-haired cat generically, and in some dialects is often called a "moggy". A horse of unknown bloodlines is a grade horse.

## Animal Feed

Animal feed is food given to domestic animals in the course of animal husbandry. There are two basic types: *fodder* and *forage*. Used alone, the word "feed" more often refers to fodder.

A photo of a feedlot in Texas, USA, where cattle are "finished" (fattened on grains) prior to slaughter.

# Fodder

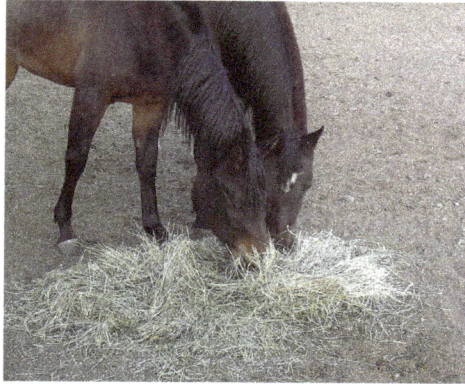

Equine nutritionists recommend that 50% or more of a horse's
diet by weight should be forages, such as hay.

"Fodder" refers particularly to foods or forages given to the animals (including plants cut and carried to them), rather than that which they forage for themselves. It includes hay, straw, silage, compressed and pelleted feeds, oils and mixed rations, and sprouted grains and legumes. Feed grains are the most important source of animal feed globally. The amount of grain used to produce the same unit of meat varies substantially. According to an estimate reported by the BBC in 2008, "Cows and sheep need 8kg of grain for every 1kg of meat they produce, pigs about 4kg. The most efficient poultry units need a mere 1.6kg of feed to produce 1kg of chicken." Farmed fish can also be fed on grain, and use even less than poultry. The two most important feed grains are maize and soybean, and the United States is by far the largest exporter of both, averaging about half of the global maize trade and 40% of the global soya trade in the years leading up the 2012 drought. Other feed grains include wheat, oats, barley, and rice, among many others.

Traditional sources of animal feed include household food scraps and the byproducts of food processing industries such as milling and brewing. Material remaining from milling oil crops like peanuts, soy, and corn are important sources of fodder. Scraps fed to pigs are called slop, and those fed to chicken are called chicken scratch. Brewer's spent grain is a byproduct of beer making that is widely used as animal feed.

A pelleted ration designed for horses.

Compound feed is fodder that is blended from various raw materials and additives. These blends are formulated according to the specific requirements of the target animal. They are manufactured by feed compounders as *meal type, pellets* or *crumbles*. The main ingredients used in commercially prepared feed are the feed grains, which include corn, soybeans, sorghum, oats, and barley.

Compound feed may also include premixes, which may also be sold separately. Premixes are composed of microingredients such as vitamins, minerals, chemical preservatives, antibiotics, fermentation products, and other ingredients that are purchased from premix companies, usually in sacked form, for blending into commercial rations. Because of the availability of these products, a farmer who uses his own grain can formulate his own rations and be assured his animals are getting the recommended levels of minerals and vitamins, although he is still subject to the Veterinary Feed Directive.

According to the American Feed Industry Association, as much as $20 billion worth of feed ingredients are purchased each year. These products range from grain mixes to orange rinds to beet pulps. The feed industry is one of the most competitive businesses in the agricultural sector, and is by far the largest purchaser of U.S. corn, feed grains, and soybean meal. Tens of thousands of farmers with feed mills on their own farms are able to compete with huge conglomerates with national distribution. Feed crops generated $23.2 billion in cash receipts on U.S. farms in 2001. At the same time, farmers spent a total of $24.5 billion on feed that year.

In 2011, around 734.5 million tons of feed were produced annually around the world.

## Forage

In this photo a herdsman from the Maasai people watches
as his cattle graze in the Ngorongoro crater, Tanzania.

"Forage" is plant material (mainly plant leaves and stems) eaten by grazing livestock. Historically, the term *forage* has meant only plants eaten by the animals directly as pasture, crop residue, or immature cereal crops, but it is also used more loosely to include similar plants cut for fodder and carried to the animals, especially as hay or silage.

## Nutrition

In agriculture today, the nutritional needs of farm animals are well understood and may be satisfied through natural forage and fodder alone, or augmented by direct supplementation of nutrients in concentrated, controlled form. The nutritional quality of feed is influenced not only by the nutrient content, but also by many other factors such as feed presentation, hygiene, digestibility, and effect on intestinal health.

Feed additives provide a mechanism through which these nutrient deficiencies can be resolved effect the rate of growth of such animals and also their health and well-being. Even with all the benefits of higher quality feed, most of a farm animal's diet still consists of grain-based ingredients because of the higher costs of quality feed.

# Feed Manufacturing

Feed manufacturing refers to the process of producing animal feed from raw agricultural products. Fodder produced by manufacturing is formulated to meet specific animal nutrition requirements for different species of animals at different life stages.

Commercial fish feed production in Stokmarknes, Norway.

# Feed and Types of Feed

The Washington State Department of Agriculture defines feed as a mix of whole or processed grains, concentrates, and commercial feeds for all species of animals to include customer formula and labeled feeds, and pet feed. These feed are now commercially produced for the livestock, poultry, swine, and fish industries. The commercial production of feed is governed by state and national laws. For example, in Texas, whole or processed grains, concentrates, and commercial feeds with the purpose of feeding wildlife and pets should be duly described in words or animation for distribution by sellers. Most State and Federal codes have clearly stated that commercial feeds should not be adulterated. Animal feeds have been broadly classified as follows:

- Concentrates: High in energy, contains mainly cereal grains and their byproducts, or is prepared from high-protein oil meals or cakes, and byproducts resulting from sugar beets and sugarcane processing.

- Roughages: Grass pastures, or plant parts like hay, silage, root crops, straw, and stover. Diets given to different species are all not the same. For example, livestock animals are fed on a diet that consists mainly of roughages, while poultry, swine, and fish are fed with concentrates. Livestock in a feedlot may be fed with energy feeds which usually comes from grains, supplied alone or as part of a total mixed ration.

# Feed Preparation and Quality

The quality of the prepared feed ultimately depends on the quality of the material such as the grain or grass used; the raw material should be of very good quality. Commercial feed manufacturing is an industrial process, and therefore should follow HACCP procedures. The Food and Drug Administration (FDA) defines HACCP as "a management system in which food safety is addressed through the analysis and control of biological, chemical, and physical hazards from raw material production, procurement and handling, to manufacturing, distribution and consumption of the finished product". The FDA regulates human food and animal feed for poultry, livestock, swine, and fish. Additionally, the FDA regulates pet food, which they estimate feeds over 177 million dogs, cats, and horses in America. Similar to human foods, animal feeds must be unadulterated and wholesome, prepared under good sanitary conditions, and truthfully be labeled to provide the required information to the consumer.

## Feed Formulation for Swine

Feed makes up approximately 60% to 80% of the total cost of producing hogs. Manufactured feeds are not merely for satiety but also must provide animals the nutrients required for healthy growth. Formulating a swine ration considers the required nutrients at various growth stages in creating an appropriate feed. Three basic methods are used to formulate swine diets: Pearson square, algebraic equations and linear programs (computers). In recent times, microcomputer programs are available that will balance a diet for many nutrients and assist with economic decisions.

The basic nutrients required are crude protein, metabolizable energy, minerals, vitamins and water. The formulation procedure has both fixed and variable portions. Swine rations are generally based on a ground cereal grain as a carbohydrate source, soybean meal as a protein source, minerals like calcium and phosphorus are added, and vitamins. The feed can be fortified with byproducts of milk, meat by-products, cereal grains; and "specialty products." Antibiotics may also be added to fortify the feed and help the animal's health and growth.

Distillers dried grains with solubles (DDGS), which are rich in energy and protein, have been used in place of corn and soybean meal in some livestock and poultry feeds, and corn DDGS have become the most popular, economical, and widely available alternative feed ingredient for use in U.S. swine diets in all phases of production. The U.S. Grain Council reported that corn DDGS is used primarily as an energy source in swine diets because it contains approximately the same amount of digestible energy (DE) and metabolizable energy (ME) as corn, although the ME content may be slightly reduced when feeding reduced-oil DDGS. A 2007 study highlighted the recent trends in the use of DDGS, as many producers are including 20% DDGS in diets of swine in all categories. Although 20% is the recommended level of inclusion, some producers are successfully using greater inclusion rates. Inclusion rate of up to 35% DDGS has been used in diets fed to nursery pigs and finishing pigs.

## Feed Formulation for Fish

Farmed fish eat specially formulated pellet feeds containing the required nutrients for both fish health and the health of humans who eat fish. A fish feed should be nutritionally well-balanced and provide a good energy source for better growth. Commercially farmed fish are broadly classified into herbivorous fish, which eat mostly plant proteins like soy or corn, vegetable oils, minerals, and vitamins; and carnivorous fish, which are given fish oils and proteins. Carnivorous fish feed contains 30-50% fish meal and oil, but recent research suggests finding alternatives to fish meal in aquaculture diets. Among the various feeds investigated, soybean meal appears to be a better alternative to fishmeal. Soybean meal prepared for the fish industry is heavily dependent on the particle sizes contained in the feed pellets. Particle size influences feed digestibility. The particle sizes of fish pellet feed are influenced by both grain properties and the milling process. Properties of the grain include hardness and moisture content. The milling process affects particle size based on the mill equipment type used, and some properties of the mill equipment (for example corrugations, gap, speed, and energy consumption).

## Feed Formulation for Poultry

As reports have indicated, feeding make-up the major cost in raising poultry animals as birds in general require feeding more than any other animals did particularly due to their faster growth rate and high rate of productivity. Feeding efficiency is reflected on the birds' performance and its products. According to National Research Council, poultry required at least 38% components in their feed. The ration of each feed components, although differ for each different stage of birds, must include carbohydrates, fats, proteins, minerals and vitamins. Carbohydrates which is usually supply by grains including corn, wheat, barley, etc. serve as major energy source in poultry feeds. Fats usually from tallow, lard or vegetables oil are essentially required to provide important fatty acid in poultry feed for membrane integrity and hormone synthesis. Proteins are important to supply the essential amino acids for the development of body tissues like muscles, nerves, cartilage, etc. Meals from soybean, canola, and corn gluten are the major source of plant protein in poultry diets. Supplementations of minerals are often required because grains, which is the main components of commercial feed contain very little amount of those. Calcium, phosphorus, chlorine, magnesium, potassium, and sodium are required in larger amounts by poultry. Vitamins, such as vitamin A, B, C, D, E, and K on the other hand are the component that required in lower amount by poultry animals.

Fanatico reported that the easiest and popular way to feed birds are to use pelleted feeds. Aside the convenience to the farmer, pelleted feeds enable the bird to eat more at a time. In addition to that, some researchers also found the improvement of feed conversion, decreasing feed wastage, improving palatability and destroying pathogens when birds were fed with pellet feed as compared to birds fed with mash feed. Commercial manufacturing of pelleted feed usually involves series of major processes including grinding, mixing and pelleting. The produced pellets are then tested for pellet durability index (PDI) to determine its quality. To enhance good health and growth, antibiotics are often added to the pelleted feed.

Researchers have concluded that smaller particle-sized feed will improve digestion due to the increasing surface area for acid and enzyme digestion in the gastrointestinal tract. However, some researchers recently brought into the attention the necessity of coarse particle for poultry feed to complement the natural design and function of gastrointestinal tract (GIT). Hetland et al and Svihus et al. discussed that the GIT retention time decreased due to lack of gizzard function that eventually gave negative impact on live performance. Zanotto & Bellaver, compared the performance of 21 day old broilers fed with different feed particle size; 0.716 mm and 1.196 mm. They found that the subject fed with larger particle size feed showed better performance. Parsons et al., evaluating different corn particle sizes in the broiler feed found that the largest particle size (2.242 mm) gave better feed intake than the other particle sizes tested (0.781, 0.950, 1.042 and 1.109 mm). Nir et al. however argued that the development of broiler was influenced by changing particle sizes. However variation in particle size between 0.5–1 mm usually did not have any effect on the broilers. Very fine particles (<0.5 mm) may impair the broilers performance due to presence of dust that cause respiratory problems, increase water intake, feed presence in the drinkers and increase litter moisture. Chewning et al., in their recent study concluded that although fine particle sizes (0.27 mm) enhanced broilers live performance, the pelleted feed did not.

All of these data show that both fine and coarse particle sizes do have different function in the poultry feed. Appropriate proportion of these two ingredient must be used with respect to the live performance of the broilers. Xu et al. compared the performance of non-pelleted feed to pellet with fine particles and found that the addition of coarse particle improved feed conversion and body weight.

## Feed Formulation for Livestock

Livestock include beef cattle, dairy cattle, horses, goats, sheep and llamas. There is no specific requirement of feed intake for each livestock because their feed continuously varies based on the animals' age, sex, breed, environment, etc. However basic nutrient requirement of a livestock's feed must consist of protein, carbohydrates, vitamins and minerals. Dairy cattle need more energy in their feed than other type of cattle. Studies have shown that energy supplied by feed is provided by various carbohydrate sources include non-fiber carbohydrates (NFC) such as fermentable feeds or neutral detergent fiber (NDF) such as forage. Feeds with high NDF is good for rumen health, however provides less energy and vice versa. Fats are added in the livestock feed to increase energy concentration, especially when the NFC content is already too high since excessive NFC lessens the NDF fraction, affecting the rumen digestion. In ruminants, most proteins consumed are breakdown by microorganisms and the microorganism later get digested by the small intestine. The crude protein required in livestock feed should be less than 7%. Lactating ruminant especially dairy cattle require highest amount of protein, especially for milk synthesis. Minerals including calcium, phosphorus and selenium are required by livestock for maintaining growth, reproduction and bone health.

Like other animals, livestock also require appropriate proportions of fine and coarse particles in their feed. Theoretically, finer particle will be easier to digest in the rumen, however the presence of coarse particle might increase the amount of starch into small intestine thus increasing energetic efficiency. Livestock could be fed by grazing on grasslands, integrated or non-integrated with crops production. Livestock that are grown in stalls or feedlots are landless and typically fed by processed feed containing veterinary drugs, growth hormones, feed additives, or nutraceuticals to improve production effectiveness. Similarly, livestock are consuming grains as the main feed or as additional nutrient to the forage based feed. Processing grains for feed is aimed to get the easiest digestible grains to maximize starch availability, thus increasing the energy supply.

Hutjens reported that milk performance was significantly better when the cattle were fed with ground corn. Aldrich (Akey Inc.) compared digestibility of various corn particle size and distribution and conclude that to have 80% digestibility, particle size of 0.5 mm should be used (for 16 hr incubation). A research team from the University of Maryland and USDA studied the development, fermentation in rumen and starch digestion sites in lactation cow feeding on corn grain from different harvests and differently processed, and concluded that digestible, metabolizable, and heat energy were higher for high moisture corn compared to dry corn. Grinding increased DMI and resulted in increased yields of milk, protein, lactose, and solids non-fat.

## Feed Manufacturing Process

Depending on the type of feed, the manufacturing process usually start with the grinding process. Figure 1 illustrates the workflow for general feed manufacturing process. Grinding of selected raw material is to produce particle sizes to be optimally and easily accepted by the animals. Depending on the formulation, feed could contain up to 10 different components including carbohydrate, protein, vitamins, minerals and additives. The feed ration can be pelleted by proportionally homogenizing the specific compositions. Pelleting is achieved by various methods, but the most common means is by extrusion. A hygienic environment is important during the entire process of the feed production to ensure quality feed.

## Grain Milling for Feed Preparations

Corn, sorghum, wheat and barley are the most used cereals in the preparation of feed for the livestock, poultry, swine, and fish industry. Roller and hammer mills are the two types of processing equipment generally used to grind grains into smaller particle sizes. Milling cereal grains by mechanical action involves several forces like compression, shearing, crushing, cutting, friction and collision. The particle size of the ground cereal is very important in the animal feed production; smaller particle sizes increase the number of particles and the surface area per unit volume which increase access to digestive enzymes. Other benefits are increased ease of handling and easier mixing of ingredients. The average particle size is given as geometric mean diameter (GMD), expressed in mm or microns (μm) and the range of variation is described by geometric standard deviation (GSD), with a larger GSD representing lower uniformity. According to Lucas, GMD and GSD are accurate descriptors of particle size distribution when the particle size distribution is expressed as log data, and are distributed log normally. Studies have shown that grinding different grains with the same mill under similar conditions results in products with different particle sizes. The hardness of a grain sample is related to the percentage of fine particles obtained after grinding, with a higher percentage of fine particles from lower hardness grains. Rose et al. discussed that hard endosperm produces irregularly-shaped larger particles, while soft endosperm produces smaller size particles. The correlation between particle size and energy consumed is although not positive but, to obtain very fine particle sizes require higher energy which reduces the rate of production. Moreover, a very fine grind of grain has no impact on the efficiency of pelleting, nor on the power consumed during pelleting. Amerah et al. discussed the availability of more data suggesting grain particle sizes are very important in mashed diets than in pelleted diets.

## Role of Animal Husbandry in Human Welfare Management

### Dairy Products

Mammalian livestock can be used as a source of milk and dairy products such as yoghurt, cheese, butter, ice cream, etc.

### Meat

It is the production of a useful form of dietary protein and energy.

## Land Management

The grazing of livestock is sometimes used as a way to control weeds and undergrowth. For example, in areas prone to wild fires, goats and sheep are set to graze on dry shrub which reduces the risk of fires.

## Fibre

Livestock produce a range of fibre/textiles. For example, sheep and goats produce wool and deer and sheep can make leather.

## Labour

Animals such as horses, donkey and yaks can be used for mechanical energy. Prior to steam power, livestock were the only available source of non-human labour. They are still used for this purpose in many places of the world, including ploughing fields, transporting goods and military functions.

## Fertilizer

Manure can be spread on fields to increase crop yields. This is an important reason why historically, plant and animal domestication have been intimately linked. Manure is also used to make plaster for walls and floors and can be used as a fuel for fires. The blood and bones of animals are also used as fertilizer.

## Management of Farm and Farm Animals

Management is the art and science of combining ideas, facilities, processes, materials and labour to produce and market a worthwhile product or service successfully. Some of the management procedures in various animal farm systems.

## Dairy Farm Management

Dairying is the management of animals for milk and its products for human consumption. Cows, buffaloes, goats and sheep are the animals that we would expect in a dairy. Cows and buffaloes generally give more milk than goats and sheep. The yellow colour of cow milk is due to the carotene.

Buffalo milk does not contain carotene. Ghee from cow fed on an abundant green fodder is more yellow than when fed on dry food. In dairy management, the people deal with processes and systems that increase yield and improve quality of milk.

## Four Essential Methods for Livestock Improvement

These are breeding, weeding, feeding and heeding, (i) Both the male and female animals selected for breeding should be of superior quality, (ii) Weeding aims that uneconomic animals must be prevented from reproducing, (iii) Feeding is also very important for animals. Each animal should be fed on a balance ration, (iv) Heeding (pay attention to) implies good animal management and general supervision including housing care and maintenance of proper cleanliness and hygiene.

## Health Care

According to WHO 'health' is the state of complete physical, mental and social well being and not merely the absence of disease. A healthy animal eats drinks and sleeps well regularly. Therefore, good health is important.

## Suitable Environmental Conditions

Adequate ventilation, suitable temperature, sufficient light, water, air and well-drained housing accommodation should be provided.

## Resistance to Diseases

If the animal is well looked after, the resistance to diseases develops and animal is protected from the diseases.

## Regular Inspections

The above mentioned measures would of course, require regular inspections, with proper record keeping. Regular visits by veterinary doctor would be necessary.

Thus the productive potentialities of live stock are controlled by three principle factors; (i) genetic makeup (ii) nutrition and (iii) environment including the climatic conditions.

## Poultry Farm Management

The word 'poultry' is used for birds which can be raised under domestication for economic purpose. The term applies to chickens, turkeys, ducks, geese, swans, guinea fowls, pigeons, peafowls and quails. In our country, it mainly means chickens, domesticated for eggs and meat. Ducks are also domesticated but to a much less extent.

## Components Poultry Farm Management

## Selection of Disease Free and Suitable Breeds

Selection of breeds is the most important aspect. The breed should be disease free and suitable to the environmental conditions. The most common egg-type variety used for commercial production throughout the world is Single Comb White Leghorn and its various strains. The meat type stocks mainly originated from Plymouth Rock, Cornish and New Hampshire breeds of fowls.

## Brood House

Brooder house should be crowd-free, rain proof and protected from predators. It should have windows with wire mesh for adequate ventilation.

## Sanitation and Hygiene

The house should be cleaned and disinfected. Good drainage system is essential to keep the poultry yard clean.

## Care of Chicks during Brooding

On the arrival of chicks sweet water (gur 50 g/ litre) is given. The feed in the form of maize dalia should be given in the first 24-28 hours, but later on complete chick feed should be added to the feeders. Additional vitamins should be given in water during the first week.

## Feed Management

Feeding constitutes the major management concern in egg and meat production. The groups of nutrients are proteins, carbohydrates, fats, minerals and vitamins.

## Light Management

Light is essential for high egg production. 14 to 16 hours of light including daylight is required for optimum production.

## Summer Management

The birds have thick feather covering and do not have sweat glands. The birds can withstand cold, but are more sensitive to heat.

## Breeds

Table: Some breeds of Chicken are given below.

| Desi | (Indigenous Breeds) | American Breeds | English Breeds |
|------|---------------------|-----------------|----------------|
| 1. | Aseel | Plymouth Rock | Australorp |
| 2. | Chittagong or Malay | Wyandotte | Cornish |
| 3. | Ghagus | Rhode Island red | Dorking |
| 4. | Busra | Jersy black giant | Orpington |
| 5. | Tenis | New Hampshire | Sussex |
| 6. | Naked Neck | | Red Cap |
| 7. | Lolab | | |

| 8.  | Karaknath   | Asiatic (other than Indian) | Mediterranean   |
| --- | ----------- | --------------------------- | --------------- |
| 9.  | Titri       | Breeds                      | Breeds          |
| 10. | Tellicherry | Brahma                      | Leghorn         |
| 11. | Danki       | Cochin                      | Minorca         |
| 12. | Kalahasti   | Langshan                    | Ancona          |
| 13. | Gallus      |                             | Andalusian (blue) |

## Poultry Diseases and their Control

The following are some of the important diseases of the poultry (i) Viral diseases : Ranikhet disease (New Castle disease), fowl-pox, infectious bronchitis, Marck's disease, chronic respiratory diseases, duck virus enteritis, hepatitis, Bird flu. (ii) Bacterial diseases: Fowl cholera, coryza, typhoid, paratyphoid, Pollorum disease, Salmonellosis, (iii) Protozoanal diseases: Coccidiosis, Spirochaetosis (tick fever). (iv) Fungal diseases: Aspergillosis (brooder pneumonia), mycosis, Aflatoxicosis— is also fungal disease in poultry which reduces immunity and spreads through contaminated food. (v) Parasites (a) External parasites : Lice, mites ticks and fleas (b) Internal parasites: Roundworms, tapeworm, threadworms, (vi) Nutritional diseases : Avitaminosis, rickets, perosis.

An outbreak of diseases like Ranikhet, Coryza or fowl cholera can lead to the death of a large number of birds. However, these and other diseases can now be controlled by preventive measures like good management, proper nutrition, and timely vaccination of the newly born chicks. Administration of sulpha drugs and broad spectrum antibiotic treatment also helps in curing several diseases.

It is also necessary to avoid overcrowding of birds, poor ventilation and dampness in poultry houses as these favour the spread of diseases. The immediate separation of the infected birds from the healthy ones and seeking veterinary aid is recommended to check the spread of the disease and its cure.

## How to Prevent the Spread of 'Bird Flu Virus'?

Bird flu is caused by a virus H5N1. All sick birds suspected of the disease must be separated from healthy birds. There must be proper disposal of dead birds and excreta. The houses are to be disinfected. The general procedure for control of infectious diseases should be followed. Complete removal of infected flock from the farm premises must be done and new flock may be introduced. Burn or bury the dead birds that die of the disease. Discourage visitors to poultry house. They can introduce infection.

## Advantages of Poultry Farming

Poultry farming has the following advantages:

## Food

It provides eggs and meat which are highly nutritious foods. They are a rich source of animal protein, minerals, right kind of fat and vitamins (A, B and D) for good health. Unfertilized eggs are called "vegetarian eggs".

## Economic Uplift

By selling the eggs and meat of these birds, the farmers become economically better. Poultry farming provides employment to a large number of people.

## Manure

The faecal matter of birds form a rich manure which increases the fertility of soil. It increases crop yields.

## Feathers

Feathers of the birds are useful.

## Recreation

The birds of poultry are also a means of recreation. Coloured chickens give pleasure look. Cock-fighting is popular in some people.

## References

- Animal-husbandry, science: britannica.com, Retrieved 26 March, 2019

- Anitei, Stefan. "Why Are Hybrids Sterile ?". Softpedia.com. Archived from the original on 1 January 2018. Retrieved 1 May 2018

- Animal-Husbandry: ancient.eu, Retrieved 27 June, 2019

- Carre, B.; Muley, N.; Gomez, J.; Ouryt, F.X.; Lafittee, E.; Guillou, D.; Signoret, C. (2005). "Soft wheat instead of hard wheat in pelleted diets results in high starch digestibility in broiler chickens". British Poultry Science. 46: 66–74. Doi:10.1080/00071660400023847

- Animal-breeding, encyclopedias-almanacs-transcripts-and-maps, science: ncyclopedia.com, Retrieved 28 July, 2019

- "Meet 18 Designer Dog Breeds". Vetstreet.com. Archived from the original on 9 February 2018. Retrieved 1 May 2018

- Benedetti, M.P.; Sartori, J.R.; Carvalho, F.B.; Pereira, L.A.; Fascina, V.B.; Stradiotti, A.C.; Pezzato, A.; Costa, C; Ferreira, J.G. (2011). "Corn texture and particle size in broiler diets". Rev. Bras. Cienc. Avic. 13 (4): 227–234. Doi:10.1590/S1516-635X2011000400002

- Animal-husbandry-role-of-animal-husbandry-in-human-welfare-management, animals: biologydiscussion.com, Retrieved 29 August, 2019

- "An Ancient Livestock, by Barbara Lang". Alpacasincanada.com. Archived from the original on 13 February 2012. Retrieved 1 May 2018

# 4
# Animal Farming Practices

Some of the common animal farming practices are cuniculture, dairying, sheep farming, intensive animal farming and pig farming. Cuniculture refers to the breeding and raising of rabbits for wool, fur or meat while dairying involves the activities related to producing, storing and distributing milk and its products. All these diverse practices of animal farming have been carefully analyzed in this chapter.

## Dairying

Dairying is the branch of agriculture that encompasses the breeding, raising, and utilization of dairy animals, primarily cows, for the production of milk and the various dairy products processed from it.

Milk for human consumption is produced primarily by the cow and the water buffalo. The goat also is an important milk producer in China, India, and other Asian countries and in Egypt. Goat's milk is also produced in Europe and North America but, compared to cow's milk, goat's milk is relatively unimportant. Buffalo's milk is produced in commercial quantities in some countries, particularly India. Where it is produced, buffalo's milk is used in the same way as is cow's milk, and in some areas the community milk supply consists of a mixture of both.

### Dairy Herds

Dairy cows are divided into five major breeds: Ayrshire, Brown Swiss, Guernsey, Holstein–Friesian, and Jersey. There are many minor breeds, among them the Red Dane, the Dutch Belted, and the Devon. There are also dual-purpose breeds used to produce milk and meat, notably the Milking Shorthorn and the Red Polled.

Holstein-Friesian cow.

The Ayrshire breed originated in Scotland. Animals of this breed are red and white or brown and white in colour, and they are strong, vigorous, and good foragers. Ayrshire milk contains about 4.1 percent butterfat. Switzerland is the native home of the Brown Swiss. These cows are silver to dark brown in colour, with a black nose and tongue. Brown Swiss are strong and vigorous. The average fat test of the milk is 4.1 percent. The Guernsey breed originated on Guernsey Island off the coast of France. The Guernsey is fawn-coloured with clear white markings. The milk averages about 4.8 percent fat and has a deep yellow colour. The Holstein–Friesian breed originated in the Netherlands. It is black and white in colour and large in size. Holsteins give more milk than any other breed; the average butterfat content is 3.7 percent. The Jersey breed originated on the isle of Jersey, in Great Britain. Jersey cows are fawn in colour, with or without white markings. They are the smallest of the major dairy breeds, but their milk is the richest, containing on the average 5.2 percent butterfat. The protein content of milk is highest for Guernsey (3.91 percent) and Jersey (3.92 percent) and lowest for Holstein (3.23 percent).

Ayrshire cow.

## Breeding and Herd Improvement

The breeds of dairy cattle have been established by years of careful selection and mating of animals to attain desired types. Increased milk and butterfat production has been the chief objective, although the objective often has shifted to increased milk and protein production. Production per cow varies with many environmental factors, but the genetic background of the cow is extremely important.

The principles of breeding to improve production have been helpful in increasing milk production in lesser developed countries. Progress has also been made in India with cows and water buffalo.

Artificial breeding has developed into a worldwide practice. Bulls with the genetic capacity to transmit high milk-producing ability to their female offspring are kept in studs. Dairy-farmer cooperatives usually operate the studs, with artificial insemination generally used. Semen for artificial insemination may be frozen for shipment to any part of the world.

## Feeding Dairy Cattle

The dairy cow is an efficient producer of human food from roughage. This ability is attributable to a unique digestive system that consists of a four-compartment stomach capable of handling roughages not digested by human beings and other monogastric (one-stomached) animals.

Pasture is the natural feed for dairy cattle, and an abundance of good pasture provides most of the requirements of a good dairy ration. An outstanding example of grassland dairying is found in New Zealand, where cows are on pasture all year and milk production costs are at a minimum. The farmer does not need to prepare and store feed for a long winter period. Feeding a balanced ration, however, rather than grass alone, increases milk production. By 2000 the average annual production per cow in New Zealand was 8,655 pounds (3,926 kilograms) of milk, while in the U.S., where supplemental feeding is common, it was 18,204 pounds, or 8,257 kilograms. Pastures of poor quality must be supplemented with other feed, such as green crops, summer silage, or hay.

During seasons when pastures are inadequate, cows need hay, silage, and grain in sufficient amounts and balance to supply nutrient needs, and to guarantee a nutritional reserve to keep milk volume and composition from declining.

## Disease Prevention

Disease is one of the greatest problems of the dairy farm. It is a constant threat and may make removal of valuable animals from the herd necessary when they show even a possibility of disease. One study of removal of cows from a typical dairy herd showed that slightly more than one in five were removed yearly and about a third of these were lost.

Good herd management includes cleanliness, isolation of sick or injured animals, keeping premises free of hazards that might cause injury, and continuous protection against poisonous plants and other material. Certain diseases, such as tuberculosis, require injections. Others, such as mastitis, require constant treatment. For some diseases there is no known cure; slaughter of the animal is the only way to stop spread of the infection. Foot and mouth disease is the most notorious of these; severe measures have been employed by most governments in order to exclude or control this disease.

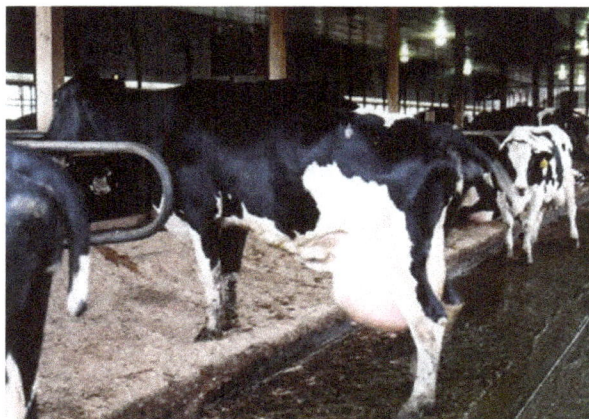

Cow suffering from mastitis.

## Milking and Bulk Handling on the Farm

The development of milk-producing tissue in the mammae is triggered by conception; minimal production begins in the seventh or eighth week, but secretion is inhibited until after calving. The stimulus of calving increases lactation for several weeks, until another conception prompts a gradual decline. In response to pregnancy hormones and the needs of the fetus, the animal is usually dry for the month or two preceding calving.

Milk is produced by the cow from her blood, and a large amount of food is necessary for maintenance of a high producing cow. The products of digestion and absorption enter the blood and are carried to the udder. There the raw materials are collected and changed into milk components. Each time the blood passes through the udder a small fraction of the components is removed to make the milk. Some 400 pounds (50 gallons, or about 200 litres) of blood must pass through the udder to make one pound (about 0.45 kilograms) of milk. A daily flow through the udder of 10 tons (20,000 pounds, or about 9,000 kilograms) of blood is required for a cow producing 50 pounds (22.5 kilograms) of milk per day. The energy required to produce milk components and to circulate the blood indicates the great importance of proper and abundant feed.

Today, most milking is done with machines by a carefully trained operator, usually twice a day, in stanchion barns or milking parlours. An experienced milker handles one to three machine units. The cows are first cleaned, and the teat cups put on. A pulsating vacuum draws the milk into a receiver or through piping into the farm milk tank.

Milk is an extremely perishable commodity that must be cooled to 50 °F (10 °C) or less within two hours. It then must be maintained at that temperature until it is delivered to the consumer.

Milk is transported from farm to plant in a variety of ways, depending on the part of the world. In the Gujarat region of India, the milk is carried to a receiving station in jars on the heads of women who do the milking. The receiving station transports the milk in large cans to the plant by truck.

In the major milk-producing countries of the world, the milk is held cold in the farm tank or in cans until it is picked up, usually once or twice daily, by tanker or truck. Tankers pump the milk in at the farm and out into plant tanks on delivery. The tanker driver measures and samples each farmer's milk; fat and bacteria tests are run at the plant. The use of pipelines has been introduced on a small scale in some European countries for delivery of milk from farm to factory.

## Sheep Farming

Sheep farming is the raising and breeding of domestic sheep. It is a branch of animal husbandry. Sheep are raised principally for their meat (lamb and mutton), milk (sheep's milk), and fiber (wool). They also yield sheepskin and parchment.

Sheep can be raised in range of temperate climates, including arid zones. Farmers build fences, housing, shearing sheds and other facilities on their property, such as for water, feed, transport and pest control. Most farms are managed so sheep can graze pastures, sometimes under the control of a shepherd or sheep dog.

Australian Merino sheep.

The major sources of income for a farm come from the sale of lambs and the shearing of sheep for their wool. Farmers can select from various breeds suitable for their region and market conditions. When the farmer sees that a ewe (female adult) is showing signs of heat or estrus, they can organise for mating with males. Newborn lambs are typically subjected to tail docking, mulesing, and males may be castrated.

Flock of sheep moving through Cologne, Germany, early on a holiday morning.

## Sheep Production Worldwide

The United Nations Food and Agriculture Organization, the top five countries by number of heads of sheep (average from 1993 to 2013) were: mainland China (146.5 million heads), Australia (101.1 million), India (62.1 million), Iran (51.7 million), and the former Sudan (46.2 million).

In 2013, the five countries with the largest number of heads of sheep were mainland China (175 million), Australia (75.5 million), India (53.8 million), the former Sudan (52.5 million), and Iran (50.2 million). In 2018 Mongolia has 30.2 million sheep. In 2013, the number of heads of sheep were distributed as follows: 44% in Asia, 28.2% in Africa; 11.2% in Europe, 9.1% in Oceania, 7.4% in the Americas.

Sheep farming in Namibia.

The top producers of sheep meat (average from 1993 to 2013) were as follows: mainland China (1.6 million); Australia (618,000), New Zealand (519,000), the United Kingdom (335,000), and Turkey (288,857). The top five producers of sheep meat in 2013 were mainland China (2 million), Australia (660,000), New Zealand (450,000), the former Sudan (325,000), and Turkey (295,000).

# Cuniculture

Cuniculture is the agricultural practice of breeding and raising domestic rabbits as livestock for their meat, fur, or wool. Cuniculture is also employed by rabbit *fanciers* and hobbyists in the development and betterment of rabbit breeds and the exhibition of those efforts. Scientists practice cuniculture in the use and management of rabbits as medical models in research. Cuniculture has been practiced all over the world since at least the 5th century.

## Aspects of Rabbit Production

## Meat Rabbits

Butchering Queensland Australia.

Rabbits have been raised for meat production in a variety of settings around the world. Smallholder or backyard operations remain common in many countries, while large-scale commercial operations are centered in Europe and Asia. For the smaller enterprise, multiple local rabbit breeds may be easier to use.

Many local, 'rustico', landrace or other heritage type breeds may be used only in a specific geographic area. Sub-par or *cull* animals from other breeding goals (laboratory, exhibition/show, wool, pet) may also be used for meat (particularly in smallholder operations).

Counterintuitively, the giant rabbit breeds are rarely used for meat production, due to their extended growth rates (which lead to high feed costs) and their large bone size (which reduces the dress-out percentage). Dwarf breeds, too, are rarely used, due to the high production costs, slow growth, and low offspring rate.

In contrast to the multitude of breeds & types used in smaller operations, breeds such as the New Zealand and the Californian, along with hybrids of these breeds, are most frequently utilized for meat in commercial rabbitries. The primary qualities of good meat-rabbit breeding stock are growth rate and size at slaughter, but also good mothering ability. Specific lines of commercial breeds have been developed that maximize these qualities – rabbits may be slaughtered as early as seven weeks and does of these strains routinely raise litters of 8 to 12 kits. Other breeds of rabbit developed for commercial meat production include the Florida White and the Altex.

A slaughtering facility in Germany (1985).

Rabbit breeding stock raised in France is particularly popular with meat rabbit farmers internationally, some being purchased as far away as in China in order to improve the local rabbit herd.

Larger-scale operations attempt to maximize income by balancing land use, labor involved, animal health, and investment in infrastructure. Specific infrastructure and strain qualities depend on the geographic area. An operation in an urban area may emphasize odor control and space utilization by stacking cages over each other with automatic cleaning systems that flush away feces and urine. In rural sub-tropical and tropical areas, temperature control becomes more of an issue, and the use of air-conditioned buildings is common in many areas.

Breeding schedules for rabbits vary by individual operation. Prior to the development of modern balanced rabbit rations, rabbit breeding was limited by the nutrition available to the doe. Without adequate calories and protein, the doe would either not be fertile, would abort or re-adsorb the foetuses during pregnancy, or would deliver small numbers of weak kits. Under these conditions, a doe would be re-bred only after weaning her last litter when the kits reached the age of two months. This allowed for a maximum of four litters per year. Advances in nutrition, such as those published by the USDA Rabbit Research Station, resulted in greater health for breeding animals and the survival of young stock. Likewise, offering superior, balanced nutrition to growing kits allowed for better health and less illness among slaughter animals. Current practices include the option of re-breeding the doe within a few days of delivery (closely matching the behavior of wild rabbits during the spring/early summer, when forage availability is at its peak.) This can result in up to eight or more litters annually. A doe of ideal meat-stock genetics can produce five times her body weight in fryers a year. Criticism of the more intensive breeding schedules has been made, on the grounds that re-breeding that closely is excessively stressful for the doe. Determination of health effects of breeding schedules is made more difficult by the domestic rabbit's reproductive psychology – in contrast to several other mammal species, rabbits are more likely to develop uterine cancer when not used for breeding than when bred frequently.

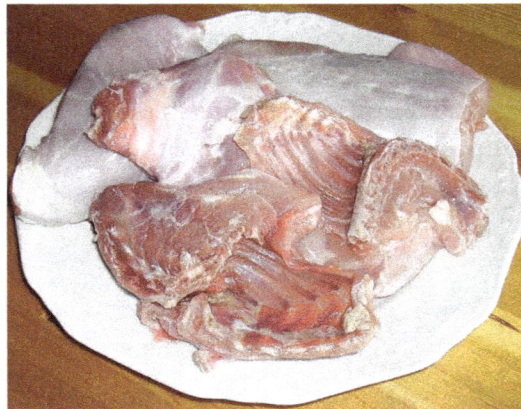

Commercially processed lean rabbit meat.

In efficient production systems, rabbits can turn 20 percent of the proteins they eat into edible meat, compared to 22 to 23 percent for broiler chickens, 16 to 18 percent for pigs and 8 to 12 percent for beef; rabbit meat is more economical in terms of feed energy than beef.

"Rabbit fryers" are rabbits that are between 70 and 90 days old, weighing 3–5 pounds (1.4–2.3 kg) in live weight. "Rabbit roasters" are rabbits from 90 days to 6 months old, weighing 5–8 pounds (2.3–3.6 kg) in live weight. "Rabbit stewers" are rabbits 6 months or older, weighing over 8 pounds (3.6 kg). "Dark fryers" (i.e., any color other than white) typically garner a lower price than "white fryers" (also called "albino fryers"), because of the slightly darker tinge to the meat. (Purely pink carcasses are preferred by most consumers.) Dark fryers are also harder to de-hide than white fryers.

In the United States, white fryers garner the highest prices per pound of live weight. In Europe, however, a sizable market remains for the dark fryers that come from older and larger rabbits. In the kitchen, dark fryers are typically prepared differently from white fryers.

In 1990, the world's annual production of rabbit meat was estimated to be 1.5 million tons. In 2014, the number was estimated at 2 million tons. China is among the world's largest producers and consumers of rabbit meat, accounting for some 30% of the world's total consumption. Within China itself, rabbits are raised in many provinces, with most of the rabbit meat (about 70% of the national production, equaling some 420,000 tons annually) being consumed in the Sichuan Basin (Sichuan Province and Chongqing), where it is particularly popular.

Well-known chef Mark Bittman wrote that domesticated rabbit "tastes like chicken", because both are "blank palettes on which we can layer whatever flavors we like.

## Wool Rabbits and Pelt Rabbits

### Wool Rabbits

Rabbits such as the Angora, American Fuzzy Lop, and Jerseys. J Wooly produce wool. However, since the American Fuzzy Lop and Jersey Wooly are both dwarf breeds, only the much larger Angora breeds such as the English Angora, Satin Angora, Giant Angora, and French Angoras are used for commercial wool production. Their long fur is sheared, combed, or plucked (gently pulling loose hairs from the body during molting) and then spun into yarn used to make a variety of products. Angora sweaters can be purchased in many clothing stores and is generally mixed with other types of wool. In 2010, 70% of Angora rabbit wool was produced in China. Rabbit wool, called Angora, is 5 times warmer than sheep's wool.

### Fur Rabbits

Rabbit pelts curing.

A number of rabbit breeds have been developed with the fur trade in mind. Breeds such as the Rex, Satin, and Chinchilla are often raised for their fur. Each breed has fur characteristics and all have a wide range of colors and patterns. "Fur" rabbits are fed a diet especially balanced for fur production and the pelts are harvested when they have reached prime condition. Rabbit fur is widely used throughout the world. China imports much of its fur from Scandinavia (80%) some from North America (5%).

## Exhibition Rabbits

Many rabbit keepers breed their rabbits for competition among other purebred rabbits of the same breed. Rabbits are judged according to the standards put forth by the governing associations of the particular country. These associations, being made up of people, may be distinctly political and reflect the preferences of particular persons on the governing boards. However, as mechanisms to preserve rare breeds of rabbits, foster communication between breeders and encourage the education of the public, these organizations are invaluable. Examples include the American Rabbit Breeders Association and the British Rabbit Council.

## Laboratory Rabbits

Rabbits have been and continue to be used in laboratory work such as production of antibodies for vaccines and research of human male reproductive system toxicology. Experiments with rabbits date back to Louis Pausture's work in France in the 1800s. In 1972, around 450 000 rabbits were used for experiments in the United States, decreasing to around 240 000 in 2006. The Environmental Health Perspective, published by the National Institute of Health, states, "The rabbit is an extremely valuable model for studying the effects of chemicals or other stimuli on the male reproductive system." According to the Humane Society of the United States, rabbits are also used extensively in the study of bronchial asthma, stroke prevention treatments, cystic fibrosis, diabetes, and cancer.

Rabbit cultivation intersects with research in two ways: first, the keeping and raising of animals for testing of scientific principles. Some experiments require the keeping of several generations of animals treated with a particular drug, in order to fully appreciate the side effects of that drug. There is also the matter of breeding and raising animals for experiments. The New Zealand White is one of the most commonly used breeds for research and testing. Specific strains of the NZW have been developed, with differing resistance to disease and cancers. Additionally, some experiments call for the use of 'specific pathogen free' animals, which require specific husbandry and intensive hygiene.

Animal rights activists generally oppose animal experimentation for all purposes, and rabbits are no exception. The use of rabbits for the Draize test, which is used for, amongst other things, testing cosmetics on animals, has been cited as an example of cruelty in animal research. Albino rabbits are typically used in the Draize tests because they have less tear flow than other animals and the lack of eye pigment make the effects easier to visualize. Rabbits in captivity are uniquely subject to rabbitpox, a condition that has not been observed in the wild.

## Husbandry

Modern methods for housing domestic rabbits vary from region to region around the globe and by type of rabbit, technological & financial opportunities and constraints, intended use, number of animals kept, and the particular preferences of the owner/farmer. Various goals include maximizing number of animals per land unit (especially common in areas with high land values or small living areas) minimizing labor, reducing cost, increasing survival and health of animals, and meeting specific market requirements (such as for clean wool, or rabbits raised on pasture.) Not all of these goals are complementary. Where the keeping of rabbits has been regulated by governments, specific requirements have been put in place. Various industries also have commonly accepted practices which produce predictable results for that type of rabbit product.

## Extensive Cuniculture Practices

Extensive cuniculture refers to the practice of keeping rabbits at a lower density and a lower production level than intensive culture. Specifically as relates to rabbits, this type of production was nearly universal prior to germ theory understanding of infectious parasites (especially coccidia) and the role of nutrition in prevention of abortion and reproductive loss. The most extensive rabbit "keeping" methods would be the harvest of wild or feral rabbits for meat or fur market, such as occurred in Australia prior to the 1990s. Warren-based cuniculture is somewhat more controlled, as the animals are generally kept to a specific area and a limited amount of supplemental feeding provided. Finally, various methods of raising rabbits with pasture as the primary food source have been developed. Pasturing rabbits within a fence (but not a cage) also known as colony husbandry, has not been commonly pursued due to the high death rate from weather and predators. More commonly (but still rare in terms of absolute numbers of rabbits and practitioners) is the practice of confining the rabbits to a moveable cage with an open or slatted floor so that the rabbits can access grass but still be kept at hand and protected from weather and predators. This method of growing rabbits does not typically result in an overall reduction for the need for supplemented feed. The growing period to market weight is much longer for grass fed rather than pellet fed animals, and many producers continue to offer small amounts of complete rations over the course of the growing period. Hutches or cages for this type of husbandry are generally made of a combination of wood and metal wire, made portable enough for a person to move the rabbits daily to fresh ground, and of a size to hold a litter of 6 to 12 rabbits at the market weight of 4 to 5 pounds. Protection from sun and driving rain are important health concerns, as is durability against predator attacks and the ability to be cleaned to prevent loss from coccidious. Medical care and the use of medicated feed are less common.

## Intensive Cuniculture Practices

Rabbits being raised on pasture in moveable enclosures
at Polyface Farm Virginia USA.

Intensive cuniculture is more focused, efficient, and time-sensitive, utilizing a greater density of animals-to-space and higher turnover. The labor required to produce each harvested hide, kilogram of wool, or market fryer—and the quality thereof—may be higher or lower than for extensive methods. Successful operations raising healthy rabbits that produce durable goods run the gamut from thousands of animals to less than a dozen. Simple hutches, kitchen floors, and even natural pits may provide shelter from the elements, while the rabbits are fed from the garden or given what can be gathered as they grow to produce a community's foodstuffs and textiles. Intensive cuniculture can also be practiced in an enclosed, climate controlled barn where rows of cages house robust rabbits eating pellets and treats before a daily health inspection or weekly weight check. Veterinary specialists and biosecurity may be part of large-scale operations, while individual rabbits in smaller setups may receive better—or worse—care.

## Challenges to Successful Production

Specific challenges to the keeping of rabbits vary by specific practices. Losses from coccidiosis are much more common when rabbits are kept on the ground (such as in warrens or colonies) or on solid floors than when on wire or slat cages that keep rabbits elevated away from urine and faeces. Pastured rabbits are more subject to predator attack. Rabbits kept indoors at an appropriate temperature rarely suffer heat loss in comparison to rabbits housed outdoors in summer. At the same time, if rabbits are housed inside without adequate ventilation, respiratory disease can be a significant cause of disease and death. Production does on fodder are rarely able to raise more than 3 litters a year without heavy losses from deaths of weak kits, abortion, and re-adsorption, all related to poor nutrition and inadequate protein intake. In contrast, rabbits fed commercial pelleted diets can face losses related to low fiber intake.

## Genetics

The study of rabbit genetics is of interest to medical researchers, fanciers, and the fur and meat industries. Each of these groups has different needs for genetic information. In the biomedical research community and the pharmaceutical industry, rabbits genetics are important for producing antibodies, testing toxicity of consumer products, and in model organism research. Among rabbit fanciers and in the fiber & fur industry, the genetics of coat color and hair properties are paramount. The meat industry relies on genetics for disease resistance, feed conversion ratio, and reproduction potential.

The rabbit genome has been sequenced and is publicly available. The mitochondrial DNA has also been sequenced. In 2011, parts of the rabbit genome were re-sequenced in greater depth in order to expose variation within the genome.

## Gene Linkage Maps

Gene = du Pattern: Dutch Gene = B Color: Black (on white).

Gene = En Pattern: English
Gene = A- B- C- D- E- Color: Chestnut.

Gene = A Agouti.

The early genetic research focused on linkage distance between various gross phenotypes using linkage analysis. Between 1924 and 1941, the relationship between the c, y, b, du, En, l, r1, r2, A, dw, w, f, and br had been established (phenotype is listed below).

- c – albino

- y – yellow fat

- du – Dutch coloring

- En – English coloring

- l – angora

- r1, r2 – rex genes

- A – Agouti

- dw – dwarf gene

- w – wide intermediate-color band

- f – furless

- br – brachydactyly

The distance between these genes is as follows, enumerated by chromosome. The format is gene1—distance—gene2 –

- c – 14.4 – y – 28.4 – b

- du – 1.2 – EN – 13.1 – l

- r1 – 17.2 – r2

- A – 14.7 – dw – 15.4 – w

- f – 28.3 – br

## Color Genes

There are 11 color gene groups (or loci) in rabbits. They are: A, B, C, D, E, En, Du, P, Si, V, W. Each locus has dominant and recessive genes. In addition to the loci there are also modifiers, which modify a certain gene. These include the rufus modifiers, color intensifiers, and plus/minus (blanket/spot) modifiers. A rabbit's coat has either two pigments (pheomelanin for yellow, and eumelanin for dark brown) or no pigment (for an albino rabbit).

Within each group, the genes are listed in order of dominance, with the most dominant gene first. In parenthesis after the description is at least one example of a color that displays this gene.

Gene = c(ch2) Medium chinchilla.

Gene = e(j) Japanese brindling (harlequin).

Gene = Enen Pattern: Broken Gene = D Color: Chocolate (on white) Gene = r1, r2 Fur type: Rex.

Gene = si Silvering of the hair shaft.

Lower case are recessive and capital letters are dominant:

- "A" represents the agouti locus (multiple bands of color on the hair shaft). The genes are:

  ○ A=agouti ("wild color" or chestnut agouti, opal, chinchilla, etc.),

  ○ A(t)=tan pattern (otter, tan, silver marten),

  ○ A=self or non-agouti (black, chocolate).

- "B" represents the brown locus. The genes are:

  ○ B=black (chestnut agouti, black otter, black),

  ○ B=brown (chocolate agouti, chocolate otter, chocolate).

- "C" represents the color locus. The genes are:

  ○ C=full color (black),

  ○ C(ch3)=dark chinchilla, removes yellow pigmentation (chinchilla, silver marten),

  ○ C(ch2)=medium (light) chinchilla, Slight reduction in eumelanin creating a more sepia tone in the fur rather than black,

  ○ C(ch1)=light (pale) chinchilla (sable, sable point, smoke pearl, seal),

  ○ C(h)=color sensitive expression of color. Warmer parts of the body do not express color. Known as Himalayan, the body is white with extremities ("points") colored in black, blue, chocolate or lilac, pink eyes,

  ○ C=albino (ruby-eyed white or REW).

- "D" represents the dilution locus. This gene dilutes black to blue and chocolate to lilac:

  ○ D=dense color (chestnut agouti, black, chocolate),

  ○ D=diluted color (opal, blue or lilac).

- "E" represents the extension locus. It works with the 'A' and 'C' loci, and rufus modifiers. When it is recessive, it removes most black pigment. The genes are:

  ○ E(d)=dominant black,

  ○ E(s)=steel (black removed from tips of fur, which then appear golden or silver),

  ○ E=normal,

  ○ E(j)=Japanese brindling (harlequin), black and yellow pigment broken into patches over the body. In a broken color pattern, this results in Tricolor,

- E=most black pigment removed (agouti becomes red or orange, self-becomes tortoise).

- "En" represents the plus/minus (blanket/spot) color locus. It is incompletely dominant and results in three possible color patterns:

  - EnEn="Charlie" or a lightly marked broken with color on ears, on nose and sparsely on body,

  - Enen=Broken rabbit with roughly even distribution of color and white,

  - Enen=Solid color with no white areas.

- "Du" represents the Dutch color pattern, (the front of the face, the front part of the body, and rear paws are white, the rest of the rabbit has colored fur). The genes are:

  - Du=absence of Dutch pattern,

  - Du(d)=Dutch (dark),

  - Du(w)=Dutch (white).

- "V" represents the vienna white locus. The genes are:

  - V=normal color,

  - Vv=Vienna carrier carries blue-eyed white gene. May appear as a solid color, with snips of white on nose and/or front paws, or Dutch marked,

  - V=vienna white (blue-eyed white or BEW).

- "Si" represents the silver locus. The genes are:

  - Si=normal color,

  - Si=silver color (silver, silver fox).

- "W" represents the middle yellow-white band locus and works with the agouti gene. The genes are:

  - W=normal width of yellow band,

  - W=doubles yellow bandwidth (Otter becomes Tan, intensified red factors in Thrianta and Belgian Hare).

- "P" represents the OCA type II form of albinism, P is because it is an integral P protein mutation. The genes are:

  - P=normal color,

  - P=albinism mutation, removes eumelanin and causes pink eyes. (Will change, for example, a Chestnut Agouti into a Shadow).

# Intensive Animal Farming

Intensive animal farming or industrial livestock production, also known as factory farming, is a production approach towards farm animals in order to maximize production output, while minimizing production costs. Intensive farming refers to animal husbandry, the keeping of livestock such as cattle, poultry, and fish at higher stocking densities than is usually the case with other forms of animal agriculture—a practice typical in industrial farming by agribusinesses. The main products of this industry are meat, milk and eggs for human consumption. There are issues regarding whether factory farming is sustainable or ethical.

Confinement at high stocking density is one part of a systematic effort to produce the highest output at the lowest cost by relying on economies of scale, modern machinery, biotechnology, and global trade. There are differences in the way factory farming techniques are practiced around the world. There is a continuing debate over the benefits, risks and ethical questions of factory farming. The issues include the efficiency of food production; animal welfare; health risks and the environmental impact (e.g. agricultural pollution and climate change).

## Contemporary Animal Production

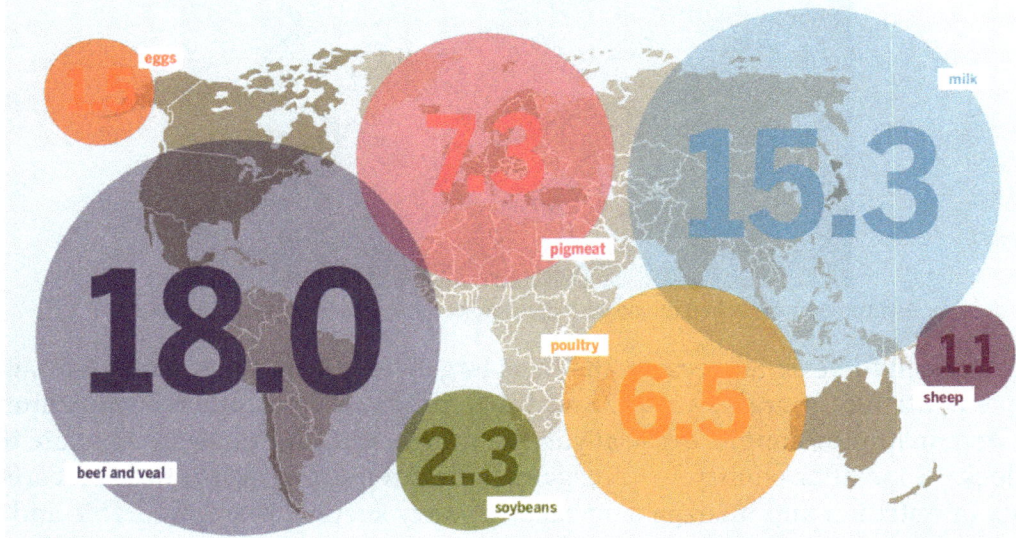

Sum of developed countries' livestock and feed subsidies.

Factory farms hold large numbers of animals, typically cows, pigs, turkeys, or chickens, often indoors, typically at high densities. The aim of the operation is to produce large quantities of meat, eggs, or milk at the lowest possible cost. Food is supplied in place. Methods employed to maintain health and improve production may include some combination of disinfectants, antimicrobial agents, anthelmintics, hormones and vaccines; protein, mineral and vitamin supplements; frequent health inspections; biosecurity; climate-controlled facilities and other measures. Physical restraints, e.g. fences or creeps, are used to control movement or actions regarded as undesirable. Breeding programs are used to produce animals more suited to the confined conditions and able to provide a consistent food product.

Intensive production of livestock and poultry is widespread in developed nations. For 2002-2003, FAO estimates of industrial production as a percentage of global production were 7 percent for beef and veal, 0.8 percent for sheep and goat meat, 42 percent for pork, and 67 percent for poultry meat. Industrial production was estimated to account for 39 percent of the sum of global production of these meats and 50 percent of total egg production. In the U.S., according to its National Pork Producers Council, 80 million of its 95 million pigs slaughtered each year are reared in industrial settings.

## Chickens

In the United States, chickens were raised primarily on family farms until 1965. Originally, the primary value in poultry was eggs, and meat was considered a byproduct of egg production. Its supply was less than the demand, and poultry was expensive. Except in hot weather, eggs can be shipped and stored without refrigeration for some time before going bad; this was important in the days before widespread refrigeration.

A commercial chicken house with open sides raising broiler pullets for meat.

Farm flocks tended to be small because the hens largely fed themselves through foraging, with some supplementation of grain, scraps, and waste products from other farm ventures. Such feedstuffs were in limited supply, especially in the winter, and this tended to regulate the size of the farm flocks. Soon after poultry keeping gained the attention of agricultural researchers, improvements in nutrition and management made poultry keeping more profitable and business like.

Prior to about 1910, chicken was served primarily on special occasions or Sunday dinner. Poultry was shipped live or killed, plucked, and packed on ice (but not eviscerated). The "whole, ready-to-cook broiler" was not popular until the 1950s, when end-to-end refrigeration and sanitary practices gave consumers more confidence. Before this, poultry were often cleaned by the neighborhood butcher, though cleaning poultry at home was a commonplace kitchen skill.

Two kinds of poultry were generally used: broilers or "spring chickens"; young male chickens, a byproduct of the egg industry, which were sold when still young and tender (generally under 3 pounds live weight), and "stewing hens", also a byproduct of the egg industry, which were old hens past their prime for laying.

Hens in Brazil.

The major milestone in 20th century poultry production was the discovery of vitamin D, which made it possible to keep chickens in confinement year-round. Before this, chickens did not thrive during the winter (due to lack of sunlight), and egg production, incubation, and meat production in the off-season were all very difficult, making poultry a seasonal and expensive proposition. Year-round production lowered costs, especially for broilers.

At the same time, egg production was increased by scientific breeding. After a few false starts, (such as the Maine Experiment Station's failure at improving egg production) success was shown by Professor Dryden at the Oregon Experiment Station. Improvements in production and quality were accompanied by lower labor requirements. In the 1930s through the early 1950s, 1,500 hens was considered to be a full-time job for a farm family. In the late 1950s, egg prices had fallen so dramatically that farmers typically tripled the number of hens they kept, putting three hens into what had been a single-bird cage or converting their floor-confinement houses from a single deck of roosts to triple-decker roosts. Not long after this, prices fell still further and large numbers of egg farmers left the business.

Robert Plamondon reports that the last family chicken farm in his part of Oregon, Rex Farms, had 30,000 layers and survived into the 1990s. However, the standard laying house of the current operators is around 125,000 hens. This fall in profitability was accompanied by a general fall in prices to the consumer, allowing poultry and eggs to lose their status as luxury foods.

The vertical integration of the egg and poultry industries was a late development, occurring after all the major technological changes had been in place for years (including the development of modern broiler rearing techniques, the adoption of the Cornish Cross broiler, the use of laying cages, etc.).

By the late 1950s, poultry production had changed dramatically. Large farms and packing plants could grow birds by the tens of thousands. Chickens could be sent to slaughterhouses for butchering and processing into prepackaged commercial products to be frozen or shipped fresh to markets or wholesalers. Meat-type chickens currently grow to market weight in six to seven weeks, whereas only fifty years ago it took three times as long. This is due to genetic selection and nutritional modifications (but not the use of growth hormones, which are illegal for use in poultry in the US and many other countries). Once a meat consumed only occasionally, the common availability and lower cost has made chicken a common meat product within developed nations. Growing concerns over the cholesterol content of red meat in the 1980s and 1990s further resulted in increased consumption of chicken.

Today, eggs are produced on large egg ranches on which environmental parameters are well controlled. Chickens are exposed to artificial light cycles to stimulate egg production year-round. In addition, forced molting is commonly practiced, in which manipulation of light and food access triggers molting, with the goal of increased egg size and production. Forced molting is controversial. While it is widespread in the US, it is prohibited in the EU.

On average, a chicken lays one egg a day, but not on every day of the year. This varies with the breed and time of year. In 1900, average egg production was 83 eggs per hen per year. In 2000, it was well over 300. In the United States, laying hens are butchered after their second egg laying season. In Europe, they are generally butchered after a single season. The laying period begins when the hen is about 18–20 weeks old (depending on breed and season). Males of the egg-type breeds have little commercial value at any age, and all those not used for breeding (roughly fifty percent of all egg-type chickens) are killed soon after hatching. The old hens also have little commercial value. Thus, the main sources of poultry meat 100 years ago (spring chickens and stewing hens) have both been entirely supplanted by meat-type broiler chickens.

Some believe that the "deadly H5N1 strain of bird flu is essentially a problem of industrial poultry practices". Transmission of highly pathogenic H5N1 from domestic poultry back to migratory waterfowl in western China has increased the geographic spread. The spread of H5N1 and its likely reintroduction to domestic poultry increase the need for good agricultural vaccines. In fact, the root cause of the continuing H5N1 pandemic threat may be the way the pathogenicity of H5N1 viruses is masked by cocirculating influenza viruses or bad agricultural vaccines.

Webster explains, In response to the same concerns, Reuters reports Hong Kong infectious disease expert Lo Wing-lok saying that "The issue of vaccines has to take top priority", and Julie Hall, in charge of the WHO's outbreak response in China, saying that China's vaccinations might be "masking" the virus. The BBC reported that Wendy Barclay, a virologist at the University of Reading, UK, said:

> "The Chinese have made a vaccine based on reverse genetics made with H5N1 antigens, and they have been using it. There has been a lot of criticism of what they have done, because they have protected their chickens against death from this virus but the chickens still get infected; and then you get drift – the virus mutates in response to the antibodies – and now we have a situation where we have five or six "flavours" of H5N1 out there".

Keeping wild birds away from domestic birds is known to be key in the fight against H5N1. Caging (no free range poultry) is one way. Providing wild birds with restored wetlands so they naturally choose nonlivestock areas is another way that helps accomplish this. Political forces are increasingly demanding the selection of one, the other, or both based on nonscientific reasons.

## Pigs

Intensive piggeries (or hog lots) are a type of concentrated animal feeding operation specialized for the raising of domestic pigs up to slaughterweight. In this system of pig production grower pigs are housed indoors in group-housing or straw-lined sheds, whilst pregnant sows are confined in sow stalls (gestation crates) and give birth in farrowing crates.

The use of sow stalls (gestation crates) has resulted in lower production costs, however, this practice has led to more significant animal welfare concerns. Many of the world's largest producers of pigs (U.S. and Canada) use sow stalls, but some nations (e.g. the UK) and some US States (e.g. Florida and Arizona) have banned them.

Intensive piggeries are generally large warehouse-like buildings. Indoor pig systems allow the pig's condition to be monitored, ensuring minimum fatalities and increased productivity. Buildings are ventilated and their temperature regulated. Most domestic pig varieties are susceptible to heat stress, and all pigs lack sweat glands and cannot cool themselves. Pigs have a limited tolerance to high temperatures and heat stress can lead to death. Maintaining a more specific temperature within the pig-tolerance range also maximizes growth and growth to feed ratio. In an intensive operation pigs will lack access to a wallow (mud), which is their natural cooling mechanism. Intensive piggeries control temperature through ventilation or drip water systems (dropping water to cool the system).

Pigs are naturally omnivorous and are generally fed a combination of grains and protein sources (soybeans, or meat and bone meal). Larger intensive pig farms may be surrounded by farmland where feed-grain crops are grown. Alternatively, piggeries are reliant on the grains industry. Pig feed may be bought packaged or mixed on-site. The intensive piggery system, where pigs are confined in individual stalls, allows each pig to be allotted a portion of feed. The individual feeding system also facilitates individual medication of pigs through feed. This has more significance to intensive farming methods, as the close proximity to other animals enables diseases to spread more rapidly. To prevent disease spreading and encourage growth, drug programs such as antibiotics, vitamins, hormones and other supplements are preemptively administered.

Indoor systems, especially stalls and pens (i.e. 'dry,' not straw-lined systems) allow for the easy collection of waste. In an indoor intensive pig farm, manure can be managed through a lagoon system or other waste-management system. However, odor remains a problem which is difficult to manage.

The way animals are housed in intensive systems varies. Breeding sows will spend the bulk of their time in sow stalls (also called gestation crates) during pregnancy or farrowing crates, with litter, until market.

Piglets often receive range of treatments including castration, tail docking to reduce tail biting, teeth clipped (to reduce injuring their mother's nipples and prevent later tusk growth) and their ears notched to assist identification. Treatments are usually made without pain killers. Weak runts may be slain shortly after birth.

Piglets also may be weaned and removed from the sows at between two and five weeks old and placed in sheds. However, grower pigs - which comprise the bulk of the herd - are usually housed in alternative indoor housing, such as batch pens. During pregnancy, the use of a stall may be preferred as it facilitates feed-management and growth control. It also prevents pig aggression (e.g. tail biting, ear biting, vulva biting, food stealing). Group pens generally require higher stockmanship skills. Such pens will usually not contain straw or other material. Alternatively, a straw-lined shed may house a larger group (i.e. not batched) in age groups.

Many countries have introduced laws to regulate treatment of farmed animals. In the USA, the federal Humane Slaughter Act requires pigs to be stunned before slaughter, although compliance and enforcement is questioned.

## Cattle

Cattle are domesticated ungulates, a member of the family Bovidae, in the subfamily Bovinae, and descended from the aurochs (*Bos primigenius*). They are raised as livestock for meat (called beef and veal), dairy products (milk), leather and as draught animals (pulling carts, plows and the like). In some countries, such as India, they are honored in religious ceremonies and revered. As of 2009–2010 it is estimated that there are 1.3–1.4 billion head of cattle in the world.

Cattle are often raised by allowing herds to graze on the grasses of large tracts of rangeland called ranches. Raising cattle in this manner allows the productive use of land that might be unsuitable for growing crops. The most common interactions with cattle involve daily feeding, cleaning and milking. Many routine husbandry practices involve ear tagging, dehorning, loading, medical operations, vaccinations and hoof care, as well as training for agricultural shows and preparations. There are also some cultural differences in working with cattle - the cattle husbandry of Fulani men rests on behavioral techniques, whereas in Europe cattle are controlled primarily by physical means like fences.

Once cattle obtain an entry-level weight, about 650 pounds (290 kg), they are transferred from the range to a feedlot to be fed a specialized animal feed which consists of corn byproducts (derived from ethanol production), barley, and other grains as well as alfalfa and cottonseed meal. The feed also contains premixes composed of microingredients such as vitamins, minerals, chemical preservatives, antibiotics, fermentation products, and other essential ingredients that are purchased from premix companies, usually in sacked form, for blending into commercial rations. Because of the availability of these products, a farmer using their own grain can formulate their own rations and be assured the animals are getting the recommended levels of minerals and vitamins.

Breeders can utilise cattle husbandry to reduce M. bovis infection susceptibility by selective breeding and maintaining herd health to avoid concurrent disease. Cattle are farmed for beef, veal, dairy, leather and they are sometimes used simply to maintain grassland for wildlife - for example, in Epping Forest, England. They are often used in some of the most wild places for livestock. Depending on the breed, cattle can survive on hill grazing, heaths, marshes, moors and semi desert. Modern cows are more commercial than older breeds and having become more specialised are less versatile. For this reason many smaller farmers still favour old breeds, such as the dairy breed of cattle Jersey.

There are many potential impacts on human health due to the modern cattle industrial agriculture system. There are concerns surrounding the antibiotics and growth hormones used, increased E. Coli contamination, higher saturated fat contents in the meat because of the feed, and also environmental concerns.

As of 2010, in the U.S. 766,350 producers participate in raising beef. The beef industry is segmented with the bulk of the producers participating in raising beef calves. Beef calves are generally raised in small herds, with over 90% of the herds having less than 100 head of cattle. Fewer producers participate in the finishing phase which often occurs in a feedlot, but nonetheless there are 82,170 feedlots in the United States.

## Criticisms

A gestational sow barn.

Advocates of factory farming claim that factory farming has led to the betterment of housing, nutrition, and disease control over the last twenty years, while opponents claim that it harms wildlife and the environment, creates health risks, abuses animals, and raises ethical issues.

In the UK, the Farm Animal Welfare Council was set up by the government to act as an independent advisor on animal welfare in 1979 and expresses its policy as five freedoms: from hunger & thirst; from discomfort; from pain, injury or disease; to express normal behavior; from fear and distress.

There are differences around the world as to which practices are accepted and there continue to be changes in regulations with animal welfare being a strong driver for increased regulation. For example, the EU is bringing in further regulation to set maximum stocking densities for meat chickens by 2010, where the UK Animal Welfare Minister commented, "The welfare of meat chickens is a major concern to people throughout the European Union. This agreement sends a strong message to the rest of the world that we care about animal welfare."

Factory farming is greatly debated throughout Australia, with many people disagreeing with the methods and ways in which the animals in factory farms are treated. Animals are often under stress from being kept in confined spaces and will attack each other. In an effort to prevent injury leading to infection, their beaks, tails and teeth are removed. Many piglets will die of shock after having their teeth and tails removed, because painkilling medicines are not used in these operations. Factory farms are a popular way to gain space, with animals such as chickens being kept in spaces smaller than an A4 page.

For example, in the UK, de-beaking of chickens is deprecated, but it is recognized that it is a method of last resort, seen as better than allowing vicious fighting and ultimately cannibalism. Between 60 and 70 percent of six million breeding sows in the U.S. are confined during pregnancy, and for most of their adult lives, in 2 by 7 ft (0.61 by 2.13 m) gestation crates. According to pork producers and many veterinarians, sows will fight if housed in pens. The largest pork producer in the U.S. said in January 2007 that it will phase out gestation crates by 2017. They are being phased out in the European Union, with a ban effective in 2013 after the fourth week of pregnancy. With the evolution of factory farming, there has been a growing awareness of the issues amongst the wider public, not least due to the efforts of animal rights and welfare campaigners. As a result, gestation crates, one of the more contentious practices, are the subject of laws in the U.S., Europe and around the world to phase out their use as a result of pressure to adopt less confined practices.

Death rates for sows have been increasing in the US from prolapse, which has been attributed to intensive breeding practices. Sows produce on average 23 piglets a year.

## Pig Farming

Pig farming is not only a profitable business but also a very popular and lucrative business. Pig is widely used to eat and pork. But it is not an easy tusk to farming. It takes a lot of time and money to make a profitable pig farm. To make a perfect pig farm everybody needs to follow some methods.

Firstly, it is very necessary to have a large area with a lot of grass and soil. Then it also needs a large fence so that the pigs can roam too far off easily and can feel comfort. But the farmers should always conscious about the structure of the fence. Farmers should use very strong wood and thick corner poles because the adult pigs become very strong and they can try to destroy the fence. As pigs are notorious diggers, farmers should dig deeply into the ground for poles. Thus farmers can farm pigs on the pasture.

Pig farming in a barn is generally easy. Farmers should use concrete to make a barn for pig farming. It is very necessary to keep the pigs inside the farm. Farmers should make the floor slope so that the water can be used to clean the barn easily. Farmers should separate the feeding area to the rest area for cleaning it easily. At least 10 feet deep and 10 feet wide area is perfect for every two pigs. The rest area should be half the size of the feeding area with a pool where they can bath and the pool should be at least 5 feet.

You should take care the pigs in the barn in the following ways:

- Depending on number of pigs farmers should clean pens naturally twice in a week.

- It is very important to keep feeder full.

- Mud hole filled with water in outside pen.

- A lot of bedding is necessary in corner but not entire pen.

## Selecting Pig Breeds

There are numerous pig breeds available throughout the world. You can choose the local breeds first which are easily available in your area. Some common and popular pig breeds are; Yorkshire, Spotted, Poland China, Landrace, Hampshire, Duroc, Chester White and Berkshire.

## Feeding

Pig should be fed 18% crude protein. Farmers should not feed pig table scraps and garbage. For this farmers may be fall in danger. To farming, the pig should give the lettuce and other vegetables, but meat products are not appropriate. If the farmers give meat products the pig will put on too much fat and it will decrease the profit. Pig should feed 10-15 pounds of feed a day when it will 130 pounds heavier. To make a pig in a perfect weight, farmers should feed ½ cup of corn oil 2 times each day mixed with 1 egg 2 times each day and ½ cup of milk re placer powder twice a day,1 table spoon corn oil is also necessary. Proper and sufficient feeding brings success in pig farming.

## Housing

To consider proper environment situation is very important to farming pig successfully. To farm pig farmers should make a pollution free environment. The pig industry needs a particular location. Without careful management of waste products, it may be very dangerous problem for child pig. It is a great and profitable process to make manure from the disposal of the pigs. It is very useful for agriculture. The environment of the outside and inside should protect all times because it is very essential to farming pigs. Inside environment is important for their health. The farming area should be made clean and dry. During the cold month, a heat lamp must be put and kept out of the north wind and south winds. Farmers should use straw as bedding during the winter. During summer season, it is very need that the pig has a place to lay in mud in the pen. If the pigs sweat or stay in a dry environment pigs will lose 5-10 pounds a day. So, always farmers should conscious about the pig's health and always try not to sweat them. To sell pigs and to get a lot of money pigs should reach at least 220-295 pounds of weight.

## Health and Diseases

Farmers should conscious about pigs' health. So medical attention is essential:

- Farmers should be aware about the symptoms of diseases. Like no interest in food, diarrhea, eye discharge, excessive coughing, hernia, dry skin and irregular spots on skin, excessively long hair, back bone showing etc.

- If the pig has diarrhea, farmers should call veterinarian and treat it carefully. Without proper treat pigs can be dehydrated and lost its weights quickly. This situation can also lead to death.

- If the pigs are coughing it is very essential to contact with veterinarian as soon as possible.

- Pigs should not give antibiotics 21-45 days before slaughter date.

- Pigs must be wormed with safe guard swine wormier or Algard monthly to kill whip.

Pig farming is very attractive and profitable business. To farm pig successfully farmer needs training and learn more things about farming. As pigs meet up the farmers economic need, it is becoming an enjoyable business day by day. To farm pigs effectively farmers should make perfect shelter for pigs. They should give proper food and proper medical protection. Without rearing pigs properly, it is not possible to earn a large number of money. So, to earn money farmers should always careful about their pigs. Though pig farming is not an easy task but it is very attractive business. To make a large farm farmers have to work a lot. Farmers have to invest huge money for farming. If any farmer follows the above instruction, then he will be surely success.

## References

- Bbott, Allison (2 August 2012). "Court Orders Temporary Closure of Dog-Breeding Facility". Nature. Doi:10.1038/nature.2012.11121

- Greenaway, Twilight (October 1, 2018). "We've bred them to their limit': death rates surge for female pigs in the US". The Guardian. Retrieved October 6, 2018

- Dairying, topic: britannica.com, Retrieved 30 June, 2019

- Lebas, Francois. "Constitution of the World Rabbit Science Association". World Rabbit Science Association. Retrieved 21 February 2018

- Pig-farming, roysfarm.com, Retrieved 29 July, 2019

- M.K. Prinsen (2006). "The Draize Eye Test and in vitro alternatives; a left-handed marriage?". Toxicology in Vitro. 20(1): Pages 78–81. Doi:10.1016/j.tiv.2005.06.030. PMID 16055303

# 5

# Livestock Management

Some of the activities which fall under the umbrella of livestock management are livestock dehorning, livestock grazing, livestock branding, livestock health and disease management, and manure management. These diverse aspects of livestock management have been thoroughly discussed in this chapter.

Livestock Production is the application of biological and chemical principles to the production and management of livestock animals and the production and handling of meat and other products. Includes instruction in animal sciences, range science, nutrition sciences, food science and technology, biochemistry, and related aspects of human and animal health and safety.

Management of livestock must take into account variable seasonal factors, fluctuating markets and declining terms of trade. The most successful producers have a good knowledge of market requirements, matching product quality to suit. There are many factors that can determine the productivity and profitability of a livestock enterprise. These include the supply and quality of feedstuffs, the use of the most appropriate genetics, ensuring high health standards, optimising housing or environmental conditions, meeting quality assurance requirements, and having a sound knowledge of market requirements. This requires good communication along the value chain.

## Livestock Dehorning

Dehorning is the removal of the horns from cattle. It is a labour-intensive, skilled operation with important animal welfare implications. Cattle can have horns of different length, shape and size, but all horns are detrimental to cattle from a welfare and production perspective, and pose a potential safety risk to cattle handlers.

Evidence indicates that the single major cause of bruising is the presence of horns.

Horned or tipped cattle (as compared to hornless cattle):

- Can cause injury to other cattle especially in yards and when in transport.
- Can cause significant damage to hides and carcase quality.
- Cause more damage to infrastructure on average.
- Need more space during transport.
- Harder to handle in yards and crushes and can be more aggressive to other animals.
- Potentially more hazardous to people.

Tipping (removal of the insensitive sharp end of the horn) is not dehorning. It does little to reduce the disadvantages of having horned cattle, for example it does not reduce bruising, and tipped cattle can still be a danger to other cattle and handlers.

## Hygiene

- Good hygiene is very important to prevent infection after dehorning.
- Dusty yards, wet days, dirty equipment, high fly activity are all factors which lead to infection.
- Equipment should be placed in a bucket of water with antiseptic, in between animals. The water should be changed frequently.
- All equipment should be cleaned thoroughly after use.

## Age

The younger the animal at the time of dehorning, the better it is for the animal and the easier the job for the operator. There is less pain and stress for the animal and there is less risk of infection or fly strike the smaller and younger they are, particularly if they are going back to their mothers. Smaller animals are also much easier to handle and restrain.

Removing horns from older cattle, yearlings and adults is time consuming, painful for the animal and increases the chances of a setback. It is not recommended to dehorn animals over 12 months of age unless undertaken by a veterinarian and is illegal in some states and territories.

## Condition of the Animal

Dehorning is not recommended when an animal is in poor condition, or if it has other health problems. In this case the animal should be dehorned when it is in better health/condition as it will heal more quickly.

## Anatomy of the Horn

The horn core is a bony extension of the skull and the hollow centre of the core communicates directly with the sinuses of the skull. The horn grows from the skin around its base, just as the wall of the hoof grows down from the skin of the coronet. To ensure no horn regrowth, it is essential to cut away 1cm of skin around the base of the horn.

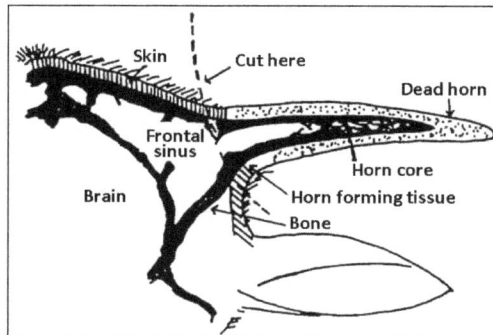

## Method

The method of dehorning should be matched to the size of the horn and the age of the animal for optimum effectiveness. The key to successful dehorning is the removal of a complete ring of hair (~1cm wide) around the horn base. The level of skill of the operator and personal preference for a particular method will also be a consideration. There is no one method for a particular animal age or horn size.

## Dehorning Knife

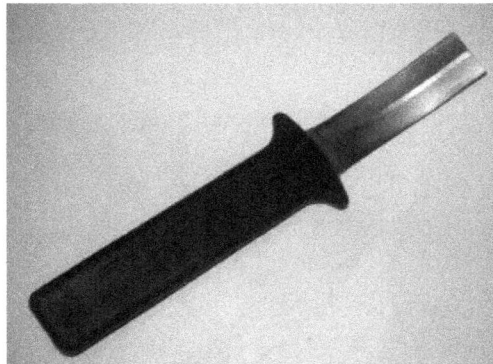

The dehorning knife has a curved blade and is easy to use. The procedure is very quick and has shown good results with no regrowth if the job is done properly. It is essential to take around 1 cm of hair around the horn bud. The knife must be kept sharp.

Dehorn calves from two weeks to about six months of age without exposing the sinus cavity (leaving a hole in their head after dehorning) as they can still have a loose, unattached horn bud. Once the bud has grown solid to the skull a small scoop dehorner may be a better choice. The dehorning knife will give excellent results with practice.

## Dehorning Iron

The use of dehorning irons does not result in an open, raw wound. Dehorning irons are made like number 'O' branding irons and can come in a variety of sizes but usually have a diameter of 50mm.

The hot iron is placed over the horn bud and the surrounding tissue and twisted a few times. This technique ensures that the blood supply will be seared off and the horn bud area will die and drop off. This technique is limited in its use when dehorning a line of animals as it is only effective on certain size horn buds. They are ideal for intermediate aged calves.

## Scoop Dehorners

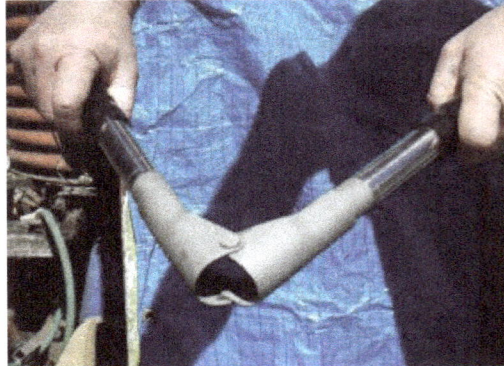

Scoop dehorners are used by passing them vertically down the horn and pushing the handles outward thereby scooping out the horn. As with other methods the head of the animal must be well restrained. Scoop dehorners come in a number of sizes and it is important the right size instrument is used on the appropriate animal. The larger size scoop dehorners can leave a deep hole if used on too smaller animal, but are useful for weaners and older cattle.

## Cup Dehorners

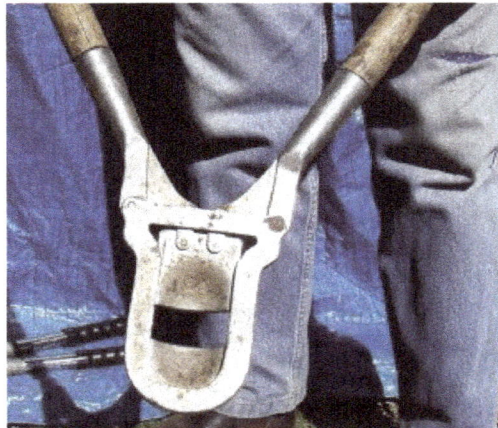

Calves that are too big to dehorn with a knife, hot iron or scoops can be dehorned with cup dehorners, which will handle cattle up to 12 months of age. Again it is necessary to take a complete ring of skin around the horn base.

## Guillotine Dehorners, Surgical Wire, Horn Saw and Tippers

Guillotine dehorners, surgical wire, horn saw and tippers are used on adult cattle with larger horn growth and should only be used for tipping.

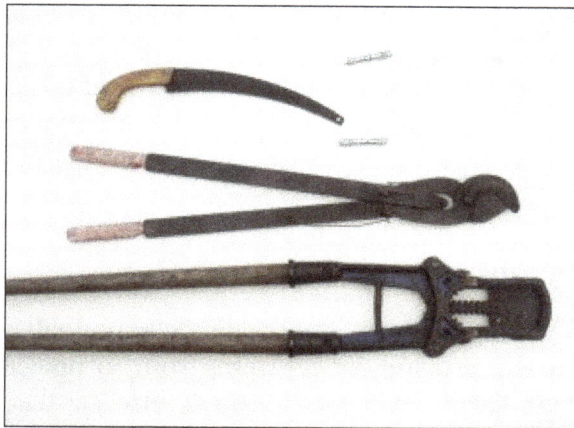

Horn saws, parrot teeth tippers and surgical wire should only be used to tip the horns of adult cattle i.e. remove only the insensitive part of the horn. If used to remove more horn, or dehorn cattle, these instruments should then only be used by a veterinarian, or under the direction of a veterinarian. Local anaesthetic should be used. Dehorning cattle over 12 months of age is not recommended and is illegal in some states.

Tipping as compared to dehorning does not reduce bruising.

## Treatment after Dehorning

- Animals bleed freely after dehorning. This is normal and helps to clean the wound.

- Fly strike is a problem when animals have an open wound but if dehorning is carried out in a hygienic manner, wounds heal up quickly. Do not apply an insecticide straight on the wound, but rather around the wound. Wound disinfectants can be applied to the wound to reduce infection risks.

- Avoid leaving animals in the yard after dehorning.

- Put dehorned stock onto good nutrition.

## Disbudding

Disbudding involves destroying the horn-producing cells (corium) of the horn bud. Horn buds are removed without opening the frontal sinus. Chemical and hot-iron disbudding methods destroy the horn-producing cells, whereas physical methods of disbudding excise them. Several methods for disbudding cattle exist, but each method has its advantages and disadvantages. Hot-iron disbudding is commonly performed and is reliable, but is considered to be quite painful. Electrical and butane hot-iron disbudding devices are available. Excessive heat applied during hot-iron disbudding can damage underlying bone. Disbudding via cautery may create less distress than physical dehorning using a scoop because nociceptors are destroyed by heat and pain perception is consequently reduced. Caustic materials (e.g., sodium hydroxide, calcium hydroxide) applied to the horn bud can damage surrounding skin and/or the eyes if runoff occurs; as long as the active chemical is in contact with tissue, damage continues. Injection of calcium chloride under the horn bud results in necrosis of the horn bud, but its administration without prior sedation and/or local anesthesia is not recommended due to the level of discomfort induced by the procedure. Cryosurgical techniques are less reliable than hot-iron disbudding, require additional procedural time, and induce behavioral indicators of pain and distress. Horn buds can be physically removed, using knives, shears, or dehorning spoons, cups, or tubes. To remove the corium and prevent horn regrowth, a complete ring of hair surrounding the horn bud should also be removed.

## Benefits of Disbudding and Dehorning

Dehorning cattle conveys advantages. Horns are the single major cause of carcass wastage due to bruising, and trim associated with bruising for carcasses from horned cattle is approximately twice that for carcasses from hornless cattle. Dehorned cattle require less feeding trough space; are easier and less dangerous to handle and transport; present a lower risk of interference from dominant animals at feeding time; pose a reduced risk of injury to udders, flanks, and eyes of other cattle; present a lower injury risk for handlers, horses, and dogs; exhibit fewer aggressive behaviors associated with individual dominance; and may incur fewer financial penalties on sale.

Disbudding of calves and kids means removing the very early developing horn base to prevent horn growth. It's a procedure carried out routinely for management reasons.

It is good practice to disbud all calves unless they are of a naturally polled type. Horns can cause a lot of damage to other cattle, and to stock handlers, particularly when they are yarded or penned or transported:

- Horned cattle should be penned separately for transport.

- There are advantages in disbudding goat kids too. Goats with horns can use them to good effect on other goats, and horns get hooked up in fences.

- Horn buds begin to appear around the time of birth or within a week or so of birth.

- Disbudding should be carried out while the buds are still very small, well before they become too large for a disbudding iron to fit over.

- Feel around the poll of young calves daily from a few days of age to check the horn buds, and disbud as soon as they form small hard caps.

- For most calves the best age for disbudding is from 3 to 6 weeks of age.

- Goat horns often appear earlier than calf horns and they grow faster, so check kids daily from birth.

## Hot Iron

- The most humane method is use of a custom-made circular hot iron to cauterize the tissue around the base of the horn.

- The procedure should take only a few seconds, but it's painful, skill is required and applying a hot iron to the head requires firm restraint of the animal.

- Don't be too forceful, especially with goat kids. Because of their smaller size and thinner skull they are more prone to injury from excess force or deep burns.

## Don't use Caustic Paste

- There are caustic chemicals on the market for disbudding.

- These are applied to the horn bud to cause chemical burns to permanently damage the horn-producing area.

- The caustic chemicals are easily rubbed onto sensitive skin (like the youngster's mother's udder or other calves!), and in wet conditions they can be washed down the face, causing painful burns.

- The risks generally don't justify use of caustic pastes for disbudding.

## Disbudding using a Scoop

- Another method of disbudding calves is by amputation using a metal scoop.

- The disbudding scoop is a special instrument designed to gouge out the small horn bud and its base.

- There is bleeding, more chance of infection than with cautery disbudding and it is a painful procedure.

- For the animal's sake, disbudding is best carried out by a veterinarian using a gas or electric cautery iron with appropriate pain control (a strong sedative, pain killer and/or anesthetic).

- The few dollars extra per calf or kid is a small price to pay for a painless and relatively stress-free procedure with a quick recovery and no complications such as infections.

- Employing a vet also means that castration, tagging and any minor surgical procedures like removal of extra teats can be carried out painlessly at the same time.

## Disbud Early

- It is much more humane to disbud calves than to dehorn older cattle. The greater size and strength of older animals make them much more difficult to restrain for dehorning, there is more bleeding and a greater risk of infection.

## Deworming

- Monthly once periodical Deworming is essential. Check the cowdung often, so that we can detect the worm infection if present. Likewise periodical deworming is reduce the Worm infection at calves and induce the growth & puberty. So that we can get high yield of weight and milk.

- Frequent checking of weight should be there to know about the growth. Then proper exercise have to provide to make the animal healthy.

## Castration

- It is very common for young male sheep, goats and calves to be castrated, because castrated animals are usually easier to manage from the age of puberty, i.e. from about 6 months of age.

- Don't castrate if you don't have to do. For example, there is no need to castrate lambs that are destined to be killed before they are 6 months old. But of course you should make sure that they are kept separate from females from about 4 months of age just in case.

- It goes without saying - castration is very painful unless it's done skillfully.

- The most humane option is to have the procedure carried out by a veterinarian using sedatives, pain killers and/or anesthetics; although many farmers consider that the costs of this make it impractical.

- Most castration of ruminants is carried out by the farmer owners, and it is best for the animal that the least stressful procedures be used and carried out while the animal is very young.

- For castration of lambs, kids and calves, the most humane method is application of a custom-made rubber ring to the neck of the scrotum with the appropriate applicator, preferable while the animal is 7 to 10 days of age and definitely before it is 6 weeks old.

- Surgical castration, ie cutting the scrotum and pulling the testicles out, is another option.

- Surgical castration is a painful operation unless it's carried out by a veterinarian using pain control.

- Keep surgical instruments clean and disinfect them between animals.

- Surgical castration is more traumatic than use of rings and there is more risk of complications like infections, so it is generally not a good method for lifestyle farmers to use.

- The risk of infections like tetanus is reduced if the mother of the animal has been fully vaccinated against clostridial diseases.

- The older the animal, the more potential there is for the operation to be painful and stressful.

- It is illegal for anyone except a veterinarian to castrate animals of any species other than sheep, goats and cattle.

## Dehorning Procedure

Dehorning can be performed on older animals and is normally performed with local anesthesia (cornual nerve block) by a veterinarian or a trained professional. Removal of larger horns is usually performed during spring and autumn to avoid fly season. Sedation may be recommended, especially for larger animals that require increased restraint. Use of longer-term pain medicine, like NSAIDs, is being researched in the US to ensure food safety.

A cow in the process of being dehorned.

For mature cattle that were not dehorned when they were young, another common practice to just cut off the pointed end of the horn. This practice is called horn tipping; it is less stressful on the animal because there is no blood loss and the horn is cut off where there are no longer any nerve endings. This practice does not eliminate the bruising damage done by the horns when cows fight, but it does eliminate the risk of puncture wounds and eye loss from pointed horns.

Disbudding minimizes discomfort and risk, and is performed when horns are small "buds" by one of several methods:

- Cauterization is the process of killing the growth ring of the horn using heat. This process is done when cattle are very young, no more than three or four weeks old—that way the horns are not very big. The earlier in the calf's life cauterization is done, the less pain and stress is inflicted on the calf. Cauterization is usually done with a dehorning hot iron after the area is numbed with local anesthesia.

- A curved knife can be used to cut the horn off when the calf is younger than a couple of months old. It is a simple procedure where the horn and the growth ring is cut off to remove the horn.

- For under eight months of age, but after the horns are starting to grow attached to the skull, so a cup dehorner or a Gigli saw, a type of surgical cutting wire, is used. There are several different types of cup dehorners, but they all serve the same function of removing the horn and growth ring. Since the horn is tougher it takes more force to remove it so tools that provide some leverage are need. Gigli saw wire is used on horns of older calf's horns that have grown too large for the cup dehorners.

- The most recent development in dehorning technology is use of a caustic dehorning paste. The paste is used on calves at less than two days old. The hair around the horn is trimmed back and then the paste is spread all over the horn bud and around the base of the horn on the growth cells. The paste kills the growth ring of the horn and then the horn falls off like a scab when it is healed. However, this method bears a risk of the paste causing injury to the animal's eyes or other tissues if used during periods of rain.

## Restraint Methods

The animal to be dehorned is usually restrained, either using a dehorning table or with chemical restraint —sedation. This ensures that the dehorning procedure can be done safely and properly. Young calves are run through a head gate (similar to a cattle crush) or haltered. Calves more than a few months old are held in a head gate and their head restrained with a dehorning table or chin bar. Smaller animals like sheep and goats may be restrained by hand or with use of halters.

## Pain Control

In 2007, the U.S. Department of Agriculture's (USDA) National Animal Health Monitoring System (NAHMS) survey suggested that most cattle in the U.S. were disbudded or dehorned without the use of anesthesia at that time. The survey showed that more than nine out of ten dairy farms practiced dehorning, but fewer than 20 percent of cattle dairy operations used analgesics or anesthesia during the process. While animal rights groups, like the Humane Society of the United States, condemn the practice of dehorning, ending it would mean increased horn-related injuries to cattle and humans. Polled genetics, long a staple in beef cattle breeding are becoming more popular among dairy farmers, with more polled calves being born to dairy cattle every year. Genetic testing can now determine if cattle carry genes for growing horns.

# Ranching

Ranching is the practice of raising herds of animals on large tracts of land. Ranchers commonly raise grazing animals such as cattle and sheep. Some ranchers also raise elk, bison, ostriches, emus, and alpacas. The ranching and livestock industry is growing faster than any other agricultural sector in the world.

Ranching is common in temperate, dry areas, such as the Pampas region of South America, the western United States, the Prairie Provinces of Canada, and the Australian Outback. In these regions, grazing animals are able to roam over large areas. Some Australian ranches, known as stations, extend more than 10,000 square kilometers (3,861 square miles). The largest, Anna Creek station, covers almost 24,000 square kilometers (9,266 square miles).

Cowboys are responsible for herding and maintaining the health of animals across these vast ranches. Cowboys often work with horses to herd cattle and sheep. Cowboy culture is an important part of the identity of ranching regions. In Mexico and South America, cowboys are known as vaqueros. In Australia and New Zealand, they are called jackaroos. Herding, round-ups, cattle drives, and branding often symbolize ranching and cowboy culture.

Herding is the practice of caring for roaming groups of livestock over a large area. Ranchers and cowboys often herd animals toward favorable grazing areas. Herding also involves keeping the herd safe from predators and natural dangers of the landscape. Grazing is so important to Australian stations, ranchers are known as graziers.

A round-up, called a muster in Australia, is a gathering of all livestock on a ranch. A round-up is usually conducted by cowboys on horseback, ATV, or other vehicle. It can be done for a wide variety of reasons: health care (such as immunization shots) for the animals, branding, or the shearing of sheep.

A round-up is one of the most difficult responsibilities of ranchers and cowboys. Animals often do not want to be rounded up and herded into a small, confined area. Even the most docile cattle or sheep are likely to become aggressive during a round-up. Round-ups also involve a large number of ranch personnel performing different tasks at the same time: veterinarians administering care to the animals, cowboys herding the animals, and wranglers caring for the ranchs horses.

A cattle drive is a massive effort of moving a herd of cattle from one place to another. In the 1700s and 1800s, cowboys on horseback took a year or more to drive cattle thousands of kilometers. Cattle drives start on ranches and usually end near points of major transportation routes, such as a harbor or railroad station. From there, cattle are loaded into vehicles and shipped to slaughterhouses.

Branding is the process of permanently marking an animal to indicate ownership. The traditional brand is known as a hot brand. A rancher or cowboy heats an iron instrument with a design unique to his ranch. Each animal belonging to that ranch has the design burned into its skin. The scar left by the burn is the animals brand.

Hot brands are less frequently used on modern ranches. Ear-tags and ink tattoos are more common. Many ranchers use microchips instead of brands. A microchip is implanted under the skin of the animal. The microchip uses radio-frequency identification (RFID) to not only identify the animals owner, but also to relay information about its location and health.

Livestock raised on ranches are an important part of a regions agriculture. Livestock provide meat for human and animal consumption. They also supply materials, such as leather and wool, for clothing, furniture, and other industries.

Some ranches, nicknamed dude ranches, offer tourist facilities. Some of these sites are working ranches that allow guests to help out in real ranching activities. Others focus on horseback riding, offering lessons and trail rides. Still others allow visitors to hunt native or imported animals. Resort ranches provide a more relaxing experience, with fun activities like trail rides and sing-alongs.

People raised livestock throughout the Middle Ages, but usually only in small numbers on small areas of land. The practice of raising large herds of livestock on extensive grazing lands started in Spain and Portugal around 1000 CE. These early ranchers used methods still associated with ranching today, such as using horses for herding, round-ups, cattle drives, and branding.

Ranching was only firmly established in the New World of the Americas. When the first Spanish explorers came to the Americas, they brought cattle and cattle-raising expertise with them. A variety of ranching traditions developed in the Americas, depending on the region the settlers came from and the characteristics of the land where they settled.

Gauchos are cowboys of the grasslands (or Pampas) of Argentina, Brazil, and Uruguay. In Central Mexico, particularly the state of Jalisco, cowboys are called charros, like the charros from Castile, Spain, who settled the region. In Northern Mexico, wealthy ranchers known as caballeros employed vaqueros to drive their cattle. Ranching in the western United States is derived from vaquero culture.

Throughout most of the 1800s, ranchers in the United States set their cattle and sheep loose to roam the prairie. Most of the grazing land was owned by the government. This was the so-called open range. Ranchers only owned enough land for a homestead and sources of water. Twice a year, cowboys rounded up cattle to brand calves (in spring) and gather steers for sale (in autumn).

Several factors contributed to the end of the open range. One was the invention of barbed wire in 1874. Farmers began to fence off their fields to protect them from being destroyed by livestock. This limited access to grazing land. Farmers and ranchers often came into conflict over land and water rights.

Overgrazing was also a problem. As more and more ranchers grazed their animals on the open range, the quality of the land became degraded. Cattle are not native to the Americas, and had to compete with native grazing animals, such as bison, for forage. Grasses did not have time to grow on the open range, especially in winter.

The winter of 1886-87, one of the harshest ever recorded, killed hundreds of thousands of cattle that were already weakened from reduced grazing. Many large ranches and cattle organizations went bankrupt. Afterwards, ranchers began fencing off their land, which they often leased from the American government.

African Americans, seeking greater freedom in the West, also worked as cowboys and ranch hands during this period.

## Working Animals

Ranches include animals other than livestock. These working animals help with the job of herding and rounding up livestock.

Horses allow cowboys to travel over rangelands quickly and keep up with moving livestock. Horses are also strong and responsive, making them excellent herding animals.

The sport of rodeo developed from the skills required of cowboys and ranch horses. Informal competitions among ranchers and cowboys tested their speed, agility, and endurance. Today, events such as roping, barrel racing, and bull riding demonstrate those same qualities among professional athletes.

Collies and sheepdogs are also used on ranches. Livestock guardian dogs do not herd animals, but are used to protect herds from predators. For example, the Great Pyrenees was bred to protect grazing animals from wolves and other predators native to the Pyrenees mountains in Spain and France.

## Ranching and the Environment

Ranching is an efficient way to raise livestock to provide meat, dairy products, and raw materials for fabrics. It is a vital part of economies and rural development around the world. However, the livestock industry has major, disruptive effects on the environment.

In South America, ranching has expanded beyond grasslands into rain forests. Ranchers clear large swaths of forest in order to create pastureland for their cattle. This clearcutting reduces habitat for native species such as monkeys, tropical birds, and millions of species of insects not found anywhere else in the world. During the past 40 years, about 20 percent of the Amazon rain forest has been cut down, much of it for cattle ranching.

Ranches established on former rain forest lands are usually not economically productive. Cleared rain forest land usually makes poor grazing land. A rain forests biodiversity exists in its above-ground canopy, not the earth beneath. Grasses do not thrive in the thin, nutrient-poor soil.

Even outside of the rain forest, many ranching practices have significant effects on the environment. Overgrazing, a threat throughout the Great Plains of the United States and Canada, puts the native tallgrass prairie ecosystem at risk. This can lead to soil erosion. The loss of valuable topsoil can reduce the agricultural productivity for crops and grazing lands. Poor agricultural practices contributed to the Dust Bowl of the 1930s, which destroyed hundreds of ranches throughout the Great Plains.

Compaction of the soil from animal hooves further degrades the land. This is unique to introduced species. Bison, native to the Americas, have small, sharp, pointed hooves. Their stampeding aerates the soil and actually contributes to the prairie ecosystem. Cattle have heavy, flat hooves that flatten the soil and reduce its ability to absorb water and nutrients.

Drylands are especially at risk for overgrazing and reduction in the quality of soil. In fact, ranching can be a key cause of desertification.

Livestock ranching also contributes to air and water pollution. Runoff from ranches can include manure, antibiotics and hormones given to the animals, as well as fertilizers and pesticides. Chemicals from tanneries that treat animal hides can also seep into water.

Ranching is also a major contributor to global warming. In fact, livestock are responsible for more greenhouse gas emissions than transportation. Carbon is released when forests are cleared for pastureland. Manure produces nitrous oxide, which has 296 times the warming potential of carbon dioxide. Cattle release large amounts of methane from their digestive systems.

Scientists, governments, and ranchers are working together to find ways to reduce these problems and make ranching a sustainable economic activity.

## Livestock Grazing Comparison

Livestock grazing comparison is a method of comparing the numbers and density of livestock grazing in agriculture. Various units of measurement are used, usually based on the grazing equivalent of one adult cow, or in some areas on that of one sheep. Many different schemes exist, giving various values to the grazing effect of different types of animal.

## Use

Livestock grazing comparison units are used for assessing the overall effect on grazing land of different types of animals (or of mixtures of animals), expressed either as a total for a whole field or farm, or as units per hectare (ha) or acre. For example, using UK government Livestock Units (LUs) from the 2003 scheme a particular 10 ha (25-acre) pasture field might be able to support 15 adult cattle or 25 horses or 100 sheep: in that scheme each of these would be regarded as being 15 LUs, or 1.5 LUs per hectare (about 0.6 LUs per acre).

Different species (and breeds) of livestock do not all graze in the same way, and this is also taken into account when deciding the appropriate number of units for grazing land. For example, horses naturally graze unevenly, eating short grass areas first and only grazing longer turf if there is insufficient short grass; cattle graze longer grass preferentially, tending to produce a uniform sward; goats tend to browse shrubs if these are available. As these feeding styles are complementary, a pasture may therefore support slightly more units of mixed species than of each species separately. Another consequence of different grazing styles is variation between species in the number of units that can lead to overgrazing – for example, horses may overgraze the short parts of a pasture even when taller grass is still available.

Livestock grazing comparison units are used by many governments to measure and control the intensity of farming. For example, until 2004 the UK Government had an extensification scheme which paid additional subsidy to farmers who kept their livestock at less than an average of 1.4 LUs per hectare.

Although different schemes have similar aims, they vary in complexity and detail. For example, some schemes give no value to a young calf, but an additional value to a cow together with her calf at foot. Some give values to different-sized animals of the same species, or different values to the same species in different regions. Most schemes use a calculation based on the weight of the animal. Some use figures for animals of different sizes which are directly proportional to their weight – for example the 2006 UK Government scheme uses a figure for ruminants of the animal's weight (in kilogrammes) divided by 650. Others include an adjustment for the proportionally higher metabolic rate of smaller animals, according to Kleiber's law, which states that the metabolic rate of most animals varies according to their weight raised to the power of approximately 0.75. For example, the Food and Agriculture Organization's Tropical Livestock Unit is based on the weight of the animal raised to the power of 0.75, compared with the equivalent figure for a "tropical cow" of 250 kg (550 lb).

The size of a livestock farm in Central Europe was traditionally given in *Stößen* (singular: *Stoß*). This unit of measurement was subsequently replaced by the grazing livestock unit or *Großvieheinheit* (GV).

## Stoß

The Stoß is a unit of cattle stock density used in the Alps. For each Alm or Alp it is worked out how many Stoß (Swiss: Stössen) can be grazed (bestoßen); one cow equals one Stoß, 3 bulls equal 2 Stöße, a calf is ¼ Stoß, a horse of 1, 2 or 3 years old is worth 1, 2 or 3 Stöße, a pig equals ¼, a goat or a sheep is ⅕ Stoß.

In Switzerland a *Normalstoß* is defined as a *Großvieheinheit* that is "summered" for 100 days. For small livestock there are corresponding conversions. Depending on the quality of the Alp or Alm a full Stoß may require between 1/2 ha and 2 ha.

The Stoß is divided into feet or Füße. A full Stoß is the pasture required by a cow, and equals 4 Füße. Bulls, calves, etc, are a fraction of that, e.g. a one-year old bull needs 2 Füße.

## Großvieheinheit

A Großvieheinheit (GV or GVE) is a conversion key used to compare different farm animals on the basis of their live weight. A Großvieheinheit represents 500 kilogrammes (roughly the weight of an adult bull). In the wild it excludes small animals like amphibians and insects, but is used for game in forestry and hunting.

Examples are:

- Calf 50–100 kg = 0.1–0.2 GV

- Young milk cow 450–650 kg = 0.9–1.3 GV

- Milk cow = 1 GV

- Horse = 0.8–1.5 GV

- Boar = 0.3 GV

- Domestic pig = 0.12 GV

- Piglet = 0.01 GV

- Sheep = 0.1 GV

- 100 Chickens = 0.8–1 GV

- 320 egg-laying chickens = 1 GV

A more precise unit is the "fodder-consuming livestock unit" or Raufutter verzehrende Großvieheinheit (RGV), which corrects the value above based on the demands of a given species and direct, near-natural supply of food (fibre-rich roughage) without concentrates.

The "tropical livestock unit" or (tropische Vieheinheit) or TLU is based on a live weight of 250 kg.

# Livestock Branding

Livestock branding is a technique for marking livestock so as to identify the owner. Originally, livestock branding only referred to hot branding large stock with a branding iron, though the term now includes alternative techniques. Other forms of livestock identification include freeze branding, inner lip or ear tattoos, earmarking, ear tagging, and radio-frequency identification (RFID), tagging with a microchip implant. The semi-permanent paint markings used to identify sheep are called a paint or colour brand. In the American West, branding evolved into a complex marking system still in use today.

A young steer is being branded with an electric branding iron and cut to make an earmark.

The act of marking livestock with fire-heated marks to identify ownership has origins in ancient times, with use dating back to the ancient Egyptians around 2,700BC. Among the ancient Romans, the symbols used for brands were sometimes chosen as part of a magic spell aimed at protecting animals from harm.

Hot iron horse branding, Spain.

Modern portable table calf branding cradle, NSW, Australia.

In English lexicon, the word "brand", common to most Germanic languages, originally meant anything hot or burning, such as a "firebrand", a burning stick. By the European Middle Ages, it commonly identified the process of burning a mark into stock animals with thick hides, such as cattle, so as to identify ownership under *animus revertendi*. The practice became particularly widespread in nations with large cattle grazing regions, such as Spain.

These European customs were imported to the Americas and were further refined by the *vaquero* tradition in what today is the southwestern United States and northern Mexico. In the American West, a "branding iron" consisted of an iron rod with a simple symbol or mark, which cowboys heated in a fire. After the branding iron turned red hot, the cowboy pressed the branding iron against the hide of the cow. The unique brand meant that cattle owned by multiple ranches could then graze freely together on the open range. Cowboys could then separate the cattle at "round-up" time for driving to market. Cattle rustlers using running irons were ingenious in changing brands. The most famous brand change involved the making of the X I T brand into the Star-Cross brand, a star with a cross inside. Brands became so numerous that it became necessary to record them in books that the ranchers could carry in their pockets. Laws were passed requiring the registration of brands, and the inspection of cattle driven through various territories. Penalties were imposed on those who failed to obtain a bill of sale with a list of brands on the animals purchased.

From the Americas, many cattle branding traditions and techniques spread to Australia, where a distinct set of traditions and techniques developed. Livestock branding has been practiced in Australia since 1866, but after 1897 owners had to register their brands. These fire and paint brands could not then be duplicated legally.

## Modern Usage

Free-range or open-range grazing is less common today than in the past. However, branding still has its uses. The main purpose is in proving ownership of lost or stolen animals. Many western US states have strict laws regarding brands, including brand registration, and require brand inspections. In many cases, a brand on an animal is considered *prima facie* proof of ownership.

In the hides and leather industry, brands are treated as a defect, and can diminish the value of hides. This industry has a number of traditional terms relating to the type of brand on a hide. "Colorado branded" (slang "Collie") refers to placement of a brand on the side of an animal, although this does not necessarily indicate the animal is from Colorado. "Butt branded" refers to a hide which has had a brand placed on the portion of the skin covering the rump area of the animal. A *cleanskin* animal is one without a brand while the skin without a brand is *native*.

Outside of the livestock industry, hot branding was used in 2003 by tortoise researchers to provide a permanent means of unique identification of individual Galapagos tortoises being studied. In this case, the brand was applied to the rear of the tortoises' shells. This technique has since been superseded by implanted PIT microchips (combined with ID numbers painted on the shell).

## Methods

The traditional cowboy or stockman captured and secured an animal for branding by roping it, laying it over on the ground, tying its legs together, and applying a branding iron that had been heated in a fire. Modern ranch practice has moved toward use of chutes where animals can be run into a confined area and safely secured while the brand is applied. Two types of restraint are the cattle crush or squeeze chute (for larger cattle), which may close on either side of a standing animal, or a branding cradle, where calves are caught in a cradle which is rotated so that the animal is lying on its side.

Bronco branding in the Top End.

Bronco branding is an old method of catching cleanskin (unbranded) cattle on Top End cattle stations for branding in Australia. A heavy horse, usually with some draught horse bloodlines and typically fitted with a harness horse collar, is used to rope the selected calf. The calf is then pulled up to several sloping topped panels and a post constructed for the purpose in the centre of the yard. The unmounted stockmen then apply leg ropes and pull it to the ground to be branded, earmarked and castrated (if a bull) there. With the advent of portable cradles, this method of branding has been mostly phased out on stations. However, there are now quite a few bronco branding competitions at rodeos and campdrafting days, etc.

Some ranches still heat branding irons in a wood or coal fire; others use an electric branding iron or electric sources to heat a traditional iron. Gas-fired branding iron heaters are quite popular in Australia, as iron temperatures can be regulated and there is not the heat of a nearby fire. Regardless of heating method, the iron is only applied for the amount of time needed to remove all hair and create a permanent mark. Branding irons are applied for a longer time to cattle than to horses, due to the differing thicknesses of their skins. If a brand is applied too long, it can damage the skin too deeply, thus requiring treatment for potential infection and longer-term healing. Branding wet stock may result in the smudging of the brand. Brand identification may be difficult on long-haired animals, and may necessitate clipping of the area to view the brand.

Horses may also be branded on their hooves, but this is not a permanent mark, so needs to be re-done about every six months. In the military, some brands indicated the horses' army and squadron numbers. These identification numbers were used on British army horses so dead horses on the battlefield could be identified. The hooves of the dead horses were then removed and returned to the Horse Guards with a request for replacements. This method was used to prevent fraudulent requests for horses. Merino rams and bulls are sometimes firebranded on their horns for permanent individual identification.

## Temporary Branding

Temporary branding is achieved by heat branding lightly, so that the hair is burned, but the skin is not damaged. Because this persists only until the animal sheds its hair, it is not considered a properly applied brand.

Some types of identification are numbering systems, neck chains, nose printing, electronic identification, and tattooing. The numbering system is a way to identify animals in a herd. It does this by putting together a letter and number to represent the year born and the birth order. The neck chains are a common way of identification with dairy cattle. The chain is labeled with a tag that has a number on it that goes along with the identification numbers. Nose printing is a common way of identification in the sale ring and at exhibiting show with some livestock. This method is like finger printing: it uses ink and cannot be modified. Electronic identification is where an electronic ear tag, microchip, or collar is placed on an animal by implanting the chip. This is done in case a tag is lost.

There are several methods of temporary branding for goats. Ear tagging, ear tattooing, and microchipping are three of these. These types of branding are usually used on goats under eight weeks of age because regular branding would harm them. Techniques similar to these are also used on sheep.

Temporary branding in ewes can be done with paint, crayons, spray markers, chalk, and much more. These can last for up to several months at a time. The sheep's identification number is painted or sprayed onto their sides or back. However, regular spray paint should never be used, as it contains chemicals that acts as painful skin irritants. Only paint that is made specifically for sheep should be put onto them.

## Freeze Branding

A white marking on the crest of a horse's neck was created by freeze branding, a form of marking for identification that is nearly painless.

Freeze brand detail on shoulder of horse.

In contrast to traditional hot-iron branding, freeze branding uses a branding iron that has been chilled with a coolant such as dry ice or liquid nitrogen. Rather than burning a scar into the animal, a freeze brand damages the pigment-producing hair cells, causing the animal's hair to grow white where the brand has been applied. Freeze brands cause less damage to the animals' hides than hot iron brands, and can be more visible. Horses are frequently freeze-branded. At this time, hogs cannot be successfully freeze branded, as their hair pigment cells are better protected. Also, freeze branding is slower, more expensive, less predictable (more care is required in application to assure desired results), and in some places does not constitute a legal brand on cattle. When an animal grows a long hair coat, the freeze brand is still visible, but its details are not always clear. Thus, it is sometimes necessary to shave or closely trim the hair so that a sharper image of a freeze brand can be viewed.

To apply a freeze brand, all hair is shaved at the branding site. This is because hair is an excellent insulator, and must be removed so the extreme cold of the freeze branding iron can be applied directly to the skin. The iron, made of metal such as brass or copper that removes heat rapidly from the skin, is submerged into the coolant. Immediately before the iron is applied, the animal's skin is rubbed, squirted, or sprayed with a generous amount of 99% alcohol, then the freeze branding iron is removed from the coolant and held onto the skin with firm pressure for several seconds. The exact amount of time will vary according to the species of the animal, the thickness of its skin, the type of metal the branding iron is made of, the type of coolant being used, and the color of its hair coat. Because a freeze-branded hair follicle regrows as white hair, a light-haired animal will have a freeze brand kept on the skin longer than does a dark-haired animal, so as to eliminate the hair follicle altogether and allow bare skin to show the brand.

Besides livestock, freeze branding can also be used on wild, hairless animals such as dolphins for purposes of tracking individuals. The brand appears as a white mark on their bare skin and can last for decades.

Immediately after the freeze branding iron is removed from the skin, an indented outline of the brand will be visible. Within seconds, however, the outline will disappear and within several minutes after that, the brand outline will reappear as swollen, puffy skin. Once the swelling subsides, for a short time, the brand will be difficult or impossible to see, but in a few days, the branded skin will begin to flake, and within three to four weeks, the brand will begin to take on its permanent appearance.

## Horse Branding Regulations

A hot brand on a Hanoverian horse together with the silhouette of the brand.

In Australia, all Arabian, Part Bred Arabians, Australian Stock Horses, Quarter Horses, Thoroughbreds, and the nine pony breeds registered in the Australian Pony Stud Book must be branded with an owner brand on the near (left) shoulder and an individual foaling drop number (in relation to the other foals) over the foaling year number on the off shoulder. In Queensland, these three brands may be placed on the near shoulder in the above order. Stock Horse and Quarter Horse classification brands are placed on the hindquarters by the classifiers.

Anglo-Arabian.   Trakehner.   Regional stud farm Moritzburg for Saxony and Thuringia.

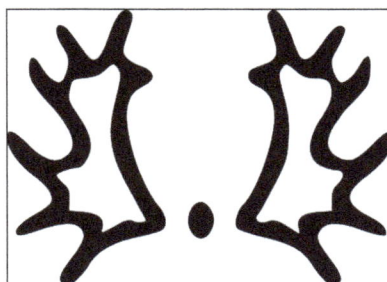

Thoroughbreds and Standardbreds in Australia and New Zealand are freeze branded. Standard-bred brands are in the form of the Alpha Angle Branding System (AABS), which the United States also uses.

Arabian.

Numerical freeze brand.

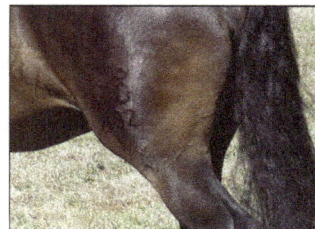
Oldenburger.

In the United States, branding of horses is not generally mandated by the government; however, there are a few exceptions: captured Mustangs made available for adoption by the BLM are freeze branded on the neck, usually with the AABS or with numbers, for identification. Horses that test positive for equine infectious anemia, that are quarantined for life rather than euthanized, will be freeze branded for permanent identification. Race horses of any breed are usually required by state racing commissions to have a lip tattoo, to be identified at the track. Some breed associations have, at times, offered freeze branding as either a requirement for registration or simply as an optional benefit to members, and individual horse owners may choose branding as a means by which to permanently identify their animals. As of 2011, the issue of whether to mandate horses be implanted with RFID microchips under the National Animal Identification System generated considerable controversy in the United States.

Bavarian Warmblood.

Holsteiner.

## Symbols and Terminology

WW I Sopwith Dolphin aircraft of No. 87 Squadron RAF used the "lazy-S" style of unit marking, derived from ranch branding.

Branding irons.

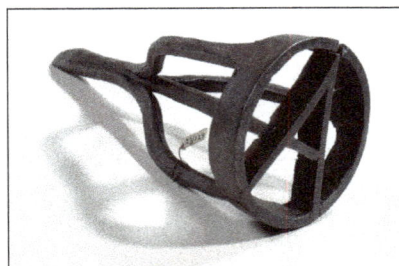

Branding iron from Swedish stallion depot.

Most brands in the United States include capital letters or numerals, often combined with other symbols such as a slash, circle, half circle, cross, or bar. Brands of this type have a specialized language for "calling" the brand. Some owners prefer to use simple pictures; these brands are called using a short description of the picture (e.g., "rising sun"). Reading a brand aloud is referred to as "calling the brand". Brands are called from left to right, top to bottom, and when one character encloses another, from outside to inside. Reading of complex brands and picture brands depends at times upon the owner's interpretation, may vary depending upon location, and it may require an expert to identify some of the more complex marks.

Terms used are:

- Upright symbols are called normally by the letters, numbers or other symbols involved.

- "Crazy": An upside down symbol. An upside down R would be read as "Crazy R",

- "Cross": A plus sign. +,

- "Slash": A forward or reverse slash. / \,

- "Rafter": Two slashes joined at the top. ∧,

- "Reverse": A reversed symbol. ⴽ would be read as "Reverse K". Reverse is sometimes called "Back" (i.e. a backwards C would be read as "Back C"),

- "Crazy Reverse": An upside down, reversed symbol. An upside down, reversed K would be read as "Crazy Reverse K",

- "Lazy": Symbols turned 90 degrees. Also, a symbol turned 90 degrees, lying on its face (or right hand side) can be read as "Lazy Down" or "Lazy Right". Similarly, a symbol turned 90 degrees, lying on its back (or left hand side) can be read as "Lazy Up" or "Lazy Left". ⌒ would be read as "Lazy 5" or "Lazy Up 5" or Lazy Left 5",

- "Tumbling": A symbol tipped about 45 degrees,

- "Flying": A symbol that starts and ends with a short serif or short horizontal line attached before the left side of the top of the symbol and attached to the right side of the symbol, extending to the right of the symbol,

- "Walking": A symbol with a short horizontal line attached to the bottom of the symbol, extending to the right of the symbol,

- "Running": A letter with a curving flare attached to the right side of the top of the letter, extending to the right, with the symbol sometimes also leaning to the right like an italic letter,

- "Over": A symbol over above another symbol, but not touching the other symbol. An H above a P would be read as "H Over P",

- "Bar": A short horizontal line. For example, a short horizontal line over an M or before an M would be read as "Bar M". Similarly, a short horizontal line under an M or after an M would be read as "M Bar". The bar can also be through the middle of the symbol and would be read as "Bar M",

- "Rail": Alternative terminology to "bar" in some areas referencing a long horizontal line. For example, a long horizontal line over a M or before an M would be read as "Rail M". Similarly, a long horizontal line under a M or after a M would be read as "M Rail",

- "Stripe": Three or more rails, one above the others,

- "Box": A symbol within a square or rectangle or a square or rectangle by itself. A box with a P inside of it would be read as "Box P",

- "Diamond": A symbol within a four sided box, the box tilted 45 degrees or a four sided box tilted 45 degrees by itself. The box sides are of equal length, and the box can be square or taller in height than in width, or greater in width than in height,

- "Rafter or Half Diamond": A half diamond over or under another symbol, but not touching the other symbol. A K with a half diamond over it, Open side facing the K, can be read as "Rafter K" or "Half Diamond K". A K with a half diamond under it, open side facing K, can be read as "K Rafter" or "K Half Diamond",

- "Circle": A symbol within a circle, or a circle by itself. A circle with a C inside of it would be read as "Circle C",

- "Half Circle or Quarter Circle": A half or quarter circle above or below a symbol, but not touching the symbol. A K with a half circle above it, open side facing up, would be read as "Half Circle K". A K with a half circle below it, open side facing down, would be read as "K Half Circle".

Combinations of symbols can be made with each symbol distinct, or:

- "Connected" or *conjoined*, with symbols touching. **TS** would be read as "T S connected" or "TS conjoined".

- "Combined or conjoined": Symbols are partially overlaid. **JK** would be read as "J K Combined".

- "Hanging": A symbol beneath another symbol and touching the other symbol. The hanging nomenclature may be omitted when reading the brand, such as a H with a P below it, with the top of the P touching the bottom of the right hand side of the H would be read as " H Hanging P", or just "H P".

- "Swinging": A symbol beneath a quarter circle, the open side of the quarter circle facing the symbol, with the symbol touching the quarter circle.For example, a H with a quarter circle over it, with the top of the H touching the quarter circle would be read as "Swinging H".

- "Rocking": A symbol above a quarter circle, the open side of the quarter circle facing the symbol, with the bottom of the symbol touching the quarter circle. For example, a H with a quarter circle under it, with the bottom of the H touching the quarter circle, is read as "Rocking H".

# Antibiotic use in Livestock

Antibiotic use in livestock is the use of antibiotics for any purpose in the husbandry of livestock, which includes treatment when ill (therapeutic), treatment of a herd of animals when at least one is diagnosed as ill (metaphylaxis), and preventative treatment (prophylaxis). The use of subtherapeutic doses in animal feed and water to promote growth and improve feed efficiency became illegal in the United States on 1 January 2017, through legislative change enacted by the Food and Drug Administration (FDA), which sought voluntary compliance from drug manufacturers to re-label their antibiotics.

## Growth Stimulation

Certain antibiotics, when given in low, subtherapeutic doses, are known to improve feed conversion efficiency (more output, such as muscle or milk, for a given amount of feed) and may promote greater growth, most likely by affecting gut flora. The drugs listed below can be used to increase feed conversion ratio and weight gain, but are not legally allowed to be used for such purposes any longer in the United States. Some drugs listed below are ionophores, which are not antibiotics and do not pose any potential risk to human health.

| Drug | Class | |
|---|---|---|
| Bacitracin | Peptide | Beef cattle, chickens, swine, and turkeys; promotes egg production in chickens |
| Bambermycin | | Beef cattle, chickens, swine, and turkeys. |
| Carbadox | | Swine |
| Laidlomycin | | Beef cattle |
| Lasalocid | Ionophore | Beef cattle |
| Lincomycin | | Chickens and swine |
| Monensin | Ionophore | Beef cattle and sheep; promotes milk production in dairy cows |
| Neomycin/ Oxytetra-cycline | | Beef cattle, chickens, swine, and turkeys |
| Penicillin | | Chickens, swine, and turkeys |
| Roxarsone | | Chickens and turkeys |
| Salinomycin | Ionophore | |
| Tylosin | | Chickens and swine |
| Virginiamycin | Peptide | Beef cattle, chickens, swine, turkeys |

## Concentrated Animal Feeding Operations

Concentrated animal feeding operations (CAFO) are large-scale, feed and housing efficient industrial facilities that raise animals in high-densities for the production of low-cost meat, eggs or milk. Most CAFOs produce only one kind of animal to improve efficiency and are classified by the type and number of animals contained, as well as how they discharge their waste water.

The classification by water discharge was implemented in 1972 when CAFOs and their smaller counterpart, AFOs, (animal feeding operations) were identified as potential sources of pollution by the Clean Water Act due to the presence of antibiotics, pathogens, and chemicals in the manure produced. The National Pollutant Discharge Elimination System (NPDES) program sets guidelines for and regulates CAFOs and AFOs.

New regulations added in 2008 under the NPDES portions of the Environmental Protection Agency (EPA) Guidelines regarding local zoning ordinances, health regulations, and Nuisance laws have still not been enforced effectively in these mass animal manufacturing operations.

## Manure

The large amounts of manure produced in cattle livestock populations are a problem for many CAFOs. Dependent on size of the operation, there can be 2,800 to 1.6 million tons of manure produced per year. Hog farms in North Carolina produce approximately 10 billion gallons of manure annually. Annually, the livestock population in the United States produces 3 to 20 times more manure than people. However, many operations lack ultimate sewage treatment.

Waste funneled from hog farms into adjacent lagoon.

The manure produced from concentrated populations of animals can be diseased and have negatives impact on the environment itself. Livestock manure can be tainted by blood, pathogens such as *E. coli*, antibiotics, growth hormones, chemical additives, etc. Some of this manure can be treated and used as fertilizer by liquefying and spraying it, but larger operations often revert to storing it until it can be disposed of properly. Large hog farms, for example, store animal waste in lagoons on site. Manure is also trucked off site, stored in containers, or held in holding ponds. There can be problems associated with storing manure; manure can have detrimental effects on the surrounding area due to leaking containers or holding ponds. This, known as *manure leaching*, can lead to manure runoff affecting the ground or soil water by percolation or direct contamination.

## Groundwater Contamination

When manure runoff or percolation enters a water system, the infecting agents thrive in that environment. Previous studies found that a private well in Idaho contained high levels of veterinary antibiotics as well as additive chemicals. The areas surrounding concentrated animal feeding operations are at particular risk for groundwater or soil water sources of contamination. When manure enters a water source, either underground or above ground, the pathogens and agents that inhabit the manure enters the water as well. Pathogens can survive longer in groundwater than surface water due to lower temperatures and protection from the sun and other harsher elements. Additionally, this water will not be treated until far later in the process, allowing bacterial colonies to grow.

The contamination of groundwater is a major concern and source of disease outbreak in humans, since it is one of the largest sources of water that humans are supplied from. As of 2015, nearly one-third of people in the United States relied on groundwater as their primary source of drinking water. Groundwater also gradually leads to surface waters, such as rivers and streams, spreading the pollution further.

## Antibiotic Usage

Pigs being housed at a CAFO.

Prior to 2017, 80% of all antibiotics were given as feed additives in the United States. While this is no longer legal, the classic administration of antibiotics in livestock is characterized by a general treatment approach where an entire herd of animals is given the same antibiotic of the same dosage, instead of treating each animal independently based on their symptoms and body size. This can lead to antibiotics not being fully synthesized by the animal and instead being excreted into the urine or manure. This has been tested by levels of antibiotics still present in the manure of these animals.

Antibiotics are also used in livestock to treat individual animals when sick, herds of animals when a few become diseased, and to prevent disease in animals that are at risk. The Centers for Disease Control and Prevention supports the responsible use of antibiotics in food animals, and the efforts the FDA and USDA are putting forth to improve antibiotic use.

## Vectors

A vector, in this context, is an organism that transmits disease to another organism. Insects such as flies and mosquitoes have high amounts of breeding grounds and nests of eggs around manure waste, allowing rapid reproduction and fresh vectors for potential disease. Typically with dense populations of livestock, transmission of disease from one animal to another can be on account of insects, such as flies, mosquitoes or ticks, spreading blood from one animal to another. This can be particularly dangerous for sick animals spreading diseases to healthier animals, promoting general malaise in a concentrated area. Additionally, the animals can be infected from other animals' manure making contact with their food; fecal-oral transmission are one of the largest sources for pathogen transmission. Within concentrated animal farming operations, there is no mandatory testing of novel viruses, only reporting known illnesses to the World Organization for Animal Health. Thus, certain mutations or recombinant bacteria strains, which are more efficient in translation to human to human events, can be unnoticed.

Additionally, insect beds around manure pools or containers are a particular threat for contamination. These insects feed and reproduce in the runoff of treated manure, so they can acquire resistant strains of bacteria from blood and the manure of livestock treated with antibiotics. Since most manure holding ponds are on or near the sites of the operations, the insects are not far from livestock populations. These insects are also particularly dangerous because they can spread bacteria or other pathogens to humans by infecting human food. This is observed when treated manure is used as fertilizers or liquefied for spraying and as a result of the unsanitary handling of meat in kitchens.

## Antibiotic Resistance

Antimicrobial resistance (AMR) can occur when antibiotics are present in concentrations too low to inhibit bacterial growth, triggering cellular responses in the bacteria that allow them to survive. These bacteria can then reproduce and spread their AMR genes to other generations, increasing their prevalence and leading to infections that cannot be healed by antibiotics.

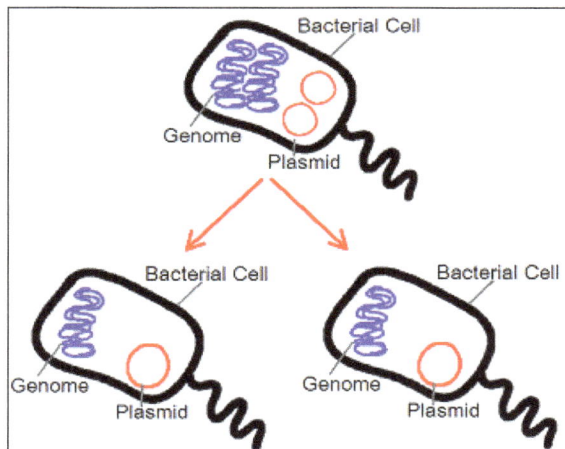

Bacterial conjugation.

Bacteria can alter their genetic inheritance through two main ways, either by mutating their genetic material or acquiring a new one from other bacteria. The latter being the most important for causing antibiotic-resistant bacteria strains in animals and humans. One of the methods bacteria can obtain new genes is through a process called conjugation which deals with transferring genes using plasmids. These conjugative plasmids carry a number of genes that can be assembled and rearranged, which could then enable bacteria to exchange beneficial genes among themselves ensuring their survival against antibiotics and rendering them ineffective to treat dangerous diseases in humans, resulting into multidrug resistant organisms.

The use of antibiotics in livestock can bring antibiotic-resistant bacteria to humans via consumption of meat and ingestion through airborne bacteria. Manure from food-producing animals can also contain antibiotic-resistant bacteria and is sometimes stored in lagoons. This waste is often sprayed as fertilizer and can thus contaminate crops and water with the bacteria.

Antibiotics are not fully digested and processed in the animal gut; therefore, an estimated 40 to 90% of the antibiotics ingested are excreted in the animal's urine and/or feces. Presence of antibiotics in animal wastes has been widely reported worldwide from multiple farms of different species. Waste from feeding organizations are usually stored on site in large lagoons to be treated before being discharged back into the environment. This composting has been shown to reduce the presence of various antibiotics presence by 20-99%. One study found that chlortetracycline (CTC), an antibiotic used in livestock feed, degraded at different rates dependent on the animal it was fed to. Half-lives of CTC were as high as 86.6 days in hog manure, while only 90% of CTC was depleted after 42 of composting for other animals. The researchers concluded that manure composting was not sufficient to ensure the microbial degradation of CTC. The high amounts of antibiotics present in animal waste, even after long periods of composting, are a potential pollution source of environmental antibiotics. Animal waste is often recycled and used as fertilizer for crops. Consequently, antibiotics, while small amounts, have been found in crops grown in these fertilized fields. In another study, antibiotics were detected in runoff from animal waste fertilized lands at concentrations up to 9 nanograms per liter. In some cases, animal wastes are directly released to receiving watersheds through intentional discharge or leaching from farms.

Overflowing lagoon.

Waste can also be released into the environment in cases of flooding. North Carolina's hog farms produces 10 billion gallons of manure annually. In 1999, Hurricane Floyd caused lagoon spillovers at 46 farms in North Carolina. Fifteen years later, a study done by the NIH showed that fecal bacteria concentrations were still high in surface waters both up and downstream from the farms. Forty percent of the samples collected exceeded state and federal guidelines for concentrations of E. coli. In preparation for the 2018 Hurricane Florence, hog farmers attempted to reduce the volume of their manure lagoons by spraying it onto surrounding fields, which, according to environmental experts, could runoff into water sources after the flooding caused by Florence. By the end of

Hurricane Florence, 26 hog manure lagoons had been damaged and 13 overflowed - meaning their contents spilled over into the floodwater.

The World Health Organization (WHO) has published a revised list in 2017, "Critically Important Antimicrobials for Human Medicine, 5th revision", with the intent that it be used "as a reference to help formulate and prioritize risk assessment and risk management strategies for containing antimicrobial resistance due to human and non-human antimicrobial use to help preserve the effectiveness of currently available antimicrobials.

## Effects in Humans

The effects of antibiotic usage in livestock transferring to humans has been well documented for over 40 years. It was first documented in 1976, where a study followed a novel antibiotic being used in livestock. The bacteria in animals and workers were regularly followed to record translational effects. The findings revealed that within 2 weeks, the bacteria found in the guts of animals fed antibiotics were resistant to the new antibiotic. Additionally, the resistant bacteria had spread to farm laborers within six months. The bacteria in the stool of the laborers were tested and contained more than 80% resistance to the initial antibiotic given to the livestock. Since the primary study, there have been many well-documented events showing that antibiotic usage in livestock results in direct influence of antibiotic resistance in humans.

In 2017, the WHO included methicillin-resistant S. aureus (MRSA) in its priority list of 12 antibiotic-resistant bacteria, urging the need to search for new and more effective antibiotics against it. There also has been an increase in the number of bacterial pathogens resistant to multiple antimicrobial agents, including MRSA, which have recently emerged into different lineages. Some of them are associated with livestock and companion animals that are then able to be transmitted to humans, also called livestock-associated methicillin-resistant Staphylococcus aureus (LA-MRSA). These new lineages have caused the infections in exposed humans to be difficult to treat and have become a public health concern.

A study looked at the association between exposure to livestock and the occurrence of livestock-associated MRSA (LA-MRSA) infection and observed that LA-MRSA infection was 9.64 times as likely to be found among livestock workers and veterinarians compared to their unexposed families and community members, showing that exposure to livestock significantly increases the risk of developing methicillin-resistant Staphylococcus aureus (MRSA) infection which has now become an increasing public health concern worldwide.

Another study found that 45.5% of industrial hog operation workers still carried in their nostrils livestock-associated Staphylococcus aureus (LA-SA) over a 14-day period up to 96h away from work, which included workers who were persistent carriers of livestock-associated methicillin-resistant Staphylococcus aureus (LA-MRSA) and livestock-associated multidrug-resistant Staphylococcus aureus (LA-MDRSA), showing that industrial hog operation workers are therefore at an increased risk in developing infections in clinical settings.

The Center for Disease Control identifies Salmonella and Campylobacter as two bacteria commonly spread to humans through food. According to the 2013 AR Threats Report, these two bacteria alone account for over 400,000 Americans becoming sick from antibiotic-resistant infections every year. When antibiotics are used for animal health, resistant bacteria can still survive and contaminate animal products during processing and slaughtering. The resistant bacteria can also be transferred to the environment through soils and water via animal manure. When the contaminated soil and water is used to fertilize crops like fruits and vegetables, then resistant bacteria is spread further to the human population.

Major bacterial infections in humans can be traced back to livestock. The family Enterobacteriaceae includes many opportunistic bacteria, including *E. coli*, which are commonly found in livestock. Other bacteria include *Klebsiella* and *Staphylococcus aureus*. They cause infections in the urinary tract, digestive system, skin, and bloodstream, and account for a significant portion of antibiotic-resistant bacterial infections. Some ways to prevent the spread of antibiotic resistant bacteria to humans is ensuring hands are clean, hands are always properly washed with soap after touching any raw foods/livestock, and are scrubbed for at least 15 seconds.

## Advocates for Restricting Antibiotic Use

Antibiotic use in animals is approved for treatment of sick animals. The FDA has approved responsible use of antibiotics to control disease spread in a population of animals when members of the population are sick, prevent spread of disease for at-risk animals, and treat infected animals. However, the CDC and FDA do not support the use of antibiotics for growth promotion because of evidence that suggests antibiotics for growth promotion lead to the development of resistant bacteria.

The practice of using antibiotics for growth stimulation is problematic for these reasons:

- It is the largest use of antimicrobials worldwide.

- Subtherapeutic use of antibiotics results in bacterial resistance.

- Every important class of antibiotics are being used in this way, making every class less effective.

- The bacteria being changed harm humans.

Donald Kennedy, former director of the FDA, has said "There's no question that routinely administering non-therapeutic doses of antibiotics to food animals contributes to antibiotic resistance." David Aaron Kessler, another former director, stated, "We have more than enough scientific evidence to justify curbing the rampant use of antibiotics for livestock, yet the food and drug industries are not only fighting proposed legislation to reduce these practices, they also oppose collecting the data."

In 2013, the US Centers for Disease Control and Prevention (CDC) published a white paper discussing antibiotic resistance threats in the US and calling for "improved use of antibiotics" among other measures to contain the threat to human health. The CDC asked leaders in agriculture, healthcare, and other disciplines to work together to combat the issue of increasing antibiotic resistance.

Some scientists have said that "all therapeutic antimicrobial agents should be available only by prescription for human and veterinary use."

The Pew Charitable Trusts have stated that "hundreds of scientific studies conducted over four decades demonstrate that feeding low doses of antibiotics to livestock breeds antibiotic-resistant superbugs that can infect people. The FDA, the U.S. Department of Agriculture and the Centers for Disease Control and Prevention all testified before Congress that there is a definitive link between the routine, non-therapeutic use of antibiotics in food animal production and the challenge of antibiotic resistance in humans."

In 2017, the World Health Organization (WHO) recommended reducing antibiotic use in animals used in the food industry. Due to the increasing risk of antibiotic resistant bacteria, the WHO strongly suggested restrictions on antibiotics being used for growth promotion and antibiotics used on healthy animals. Animals that require antibiotics should be treated with antibiotics that pose the smallest risk to human health.

HSBC produced a report in October 2018 warning that the use of antibiotics in meat production could have "devastating" consequences for humans. It noted that many dairy and meat producers in Asia and the Americas had an economic incentive to continue high usage of antibiotics, particularly in crowded or unsanitary living conditions.

## Moderate Positions

The World Organisation for Animal Health has acknowledged the need to protect antibiotics but argued against a total ban on antibiotic use in animal production.

A total ban on antibiotics might drastically reduce protein supply in some parts of the world.

## Advocates for Status Quo

In 2011 the National Pork Producers Council, an American trade association, has said, "Not only is there no scientific study linking antibiotic use in food animals to antibiotic resistance in humans, as the U.S. pork industry has continually pointed out, but there isn't even adequate data to conduct a study." The statement contradicts scientific consensus, and was issued in response to a United States Government Accountability Office report that asserts, "Antibiotic use in food animals contributes to the emergence of resistant bacteria that may affect humans".

The National Pork Board, a government-owned corporation of the United States, has said, "The vast majority of producers use (antibiotics) appropriately".

## Effects of Restricting Antibiotic Use

When government regulation restricts use of antibiotics, the negative economic impact is not often considered.

When antibiotics are used subtheraputically (for animal performance, increased growth, and improved feed efficiency), then the costs of meat, eggs, and other animal products are lowered. One big argument against the restriction of antibiotic use is the potential economic hardship that would result for producers of livestock and poultry that could also result in higher cost for consumers. In a study analysing the economic cost of the FDA restricting all antibiotic use in animal livestock, it was estimated that the restriction would cost consumers approximately $1.2 billion to $2.5 billion per year. In order to determine the overall economic impact of restricting antibiotic use, the financial cost must be weighed against the health benefits to the population. Since it is difficult to estimate quantifications of potential health benefits, the study concluded that the complete economic impact of restricting antibiotic use has not yet been determined.

Although quantifying health benefits may be difficult, the economic impact of antibiotic restriction in animals can also be evaluated through the economic impact of antibiotic resistance in humans, which is a significant outcome of antibiotic use in animals. The World Health Organization identifies antibiotic resistance as a contributor to longer hospital stays and higher medical costs. When infections can no longer be treated by typical first-line antibiotics, more expensive medications are required for treatment. When illness duration is extended by antibiotics resistance, the increased health care costs create a larger economic burden for families and societies. The Center for Infectious Disease Research and Policy estimates approximately $2.2 billion in antibiotic resistance- related healthcare costs each year.

While restricting antibiotics in animals causes a significant economic burden, the outcome of antibiotic resistance in humans that is perpetuated by antibiotic use in animals carries a comparable economic burden.

Regulation of antibiotics in livestock production would affect the business models of corporations such as Tyson Foods, Cargill, and Hormel.

## Difficulties with Determining Relevant Facts

It is difficult to set up a comprehensive surveillance system for measuring rates of change in antibiotic resistance. The US Government Accountability Office published a report in 2011 stating that government and commercial agencies had not been collecting sufficient data to make a decision about best practices.

Currently, there is no regulatory agency in the United States that systematically collects detailed data on antibiotic use in humans and animals. It is not clear which antibiotics are prescribed for which purpose and at what time. Furthermore, the world has no surveillance infrastructure to monitor emerging antibiotic resistance threats. Because of these issues, it is difficult to quantify antibiotic resistance, to regulate antibiotic prescribing practices, and to detect and respond to rising threats.

## Specific Resistance that has been Identified

At this time, the most well-documented impact on humans is foodborne gastrointestinal illness. In most cases, these illnesses are mild and do not require antibiotics; though if the infectious bacteria is drug-resistant, they have increased virulence (ability to cause disease) and lead to prolonged illness. Furthermore, in approximately 10% of cases, the disease becomes severe, requiring more advanced treatments. These treatments can take the form of intravenous antibiotics, supportive care for blood infections, and hospital stays, leading to higher costs and greater morbidity, with a trend toward higher mortality. Though all people are susceptible, populations at higher risk for severe disease include children, the elderly, and those with chronic disease.

Over the past 20 years, the most common drug-resistant foodborne bacteria in industrialized countries have been non-typhoidal *Salmonella* and *Campylobacter*. Research has consistently shown the main contributing factors are bacteria sourced in livestock. A 1998 outbreak of multi-drug-resistant *Salmonella* in Denmark linked back to two Danish swine herds. Coupled with the discovery of this link, there have been improved monitoring systems that have helped to quantify the impact. In the United States, it is estimated that there are approximately 400,000 cases and over 35,000 hospitalizations per year attributable to increasing resistant strains of *Salmonella* and *Campylobacter*. In terms of financial impact in the US, the treatment of non-typhoidal *Salmonella* infections alone is now estimated to cost $365 million per year. In light of this, in its inaugural 2013 report on antibiotic resistance threats in the United States, the CDC identified resistant non-typhoidal *Salmonella* and *Campylobacter* as "serious threats" and called for improved surveillance and intervention in food production moving forward.

There are other bacteria as well, where research is evolving and revealing that bacterial resistance acquired through use in livestock may be contributing to disease in humans. Examples of these include Enterococcus (including *E. coli* 0157) and *Staphylococcus aureus*. For foodborne illness from *E. coli*, which is still not typically treated with antibiotics because of associated risk of renal failure, increasing rates of antibiotic-resistant infections have been correlated with increasing virulence of the bacteria. In the case of Enterococcus and *S. aureus*, resistant forms of both of these bacteria have resulted in greatly increasing morbidity and mortality in the US. At this point, there have been studies, though a limited number, that definitively link antibiotic use in food production to these resistance patterns in humans and further research will help to further characterize this relationship.

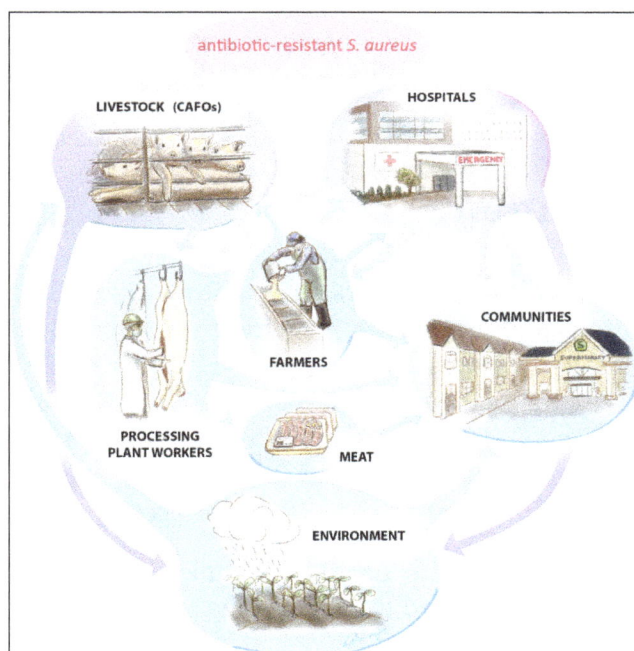

The origins of antibiotic-resistant Staphylococcus aureus (CAFO:
concentrated animal feeding operations).

## Mechanisms for Transfer to Humans

Antibiotics given in concentrations too low to combat disease are called "subtherapeutic." The administration of these drugs when there is no diagnosis of disease result in decreased mortality and morbidity and increased growth in the animals treated. It is theorized that subtherapeutic doses kill some, but not all, of the bacterial organisms in the animal – likely leaving those that are naturally antibiotic-resistant. The actual mechanism by which subtherapeutic antibiotic feed additives serve as growth promoters is still unclear. Some people have speculated that animals and fowl may have subclinical infections, which would be cured by low levels of antibiotics in feed, thereby allowing the animals to thrive.

Humans can be exposed to antibiotic-resistant bacteria by ingesting them through the food supply. Dairy products, ground beef, and poultry are the most common foods harboring these pathogens. There is evidence that a large proportion of resistant *E. coli* isolates causing bloodstream infections in people are from livestock produced as food.

When manure from antibiotic-fed swine is used as fertilizer elsewhere, the manure may be contaminated with bacteria which can infect humans.

Studies have also shown that direct contact with livestock can lead to the spread of antibiotic-resistant bacteria from animals to humans.

Sampling of retail meats such as turkey, chicken, pork and beef consistently show high levels of Enterobacteriaceae. Rates of resistant bacteria in these meats are high as well. Sources on contaminated meat put humans at direct risk by handling the meat or ingesting it before it is completely cooked. Food preservation methods can help eliminate, decrease, or prevent the growth of bacteria. Evidence for the transfer of macrolide-resistant microorganisms from animals to humans has been scant, and most evidence shows that pathogens of concern in human populations originated in humans and are maintained there, with rare cases of transference to humans.

## Manure Management

Nutrient management is particularly important in Manitoba due to the large expansion of the hog industry over the last few decades. Although manure is an ebxcellent source for plant nutrients, the expansion to more than 8 million hogs in 2010 has resulted in the challenge of too much manure and not enough land-base for spreading. The large amount of manure created and the resulting nitrous oxide ($N_2O$) should make manure management a priority when trying to reduce greenhouse gas (GHG) emissions from farms. Approximately 9 percent of Manitoba's agricultural GHG emissions are created due to manure storage and management.

Major emissions from manure come in the form of methane ($CH_4$) from anaerobic decomposition of manure during storage, and $N_2O$ formed during storage and application.

The creation of these gases is influenced by a variety of factors: temperature, oxygen level, moisture or amount of nutrients.

In turn, these factors are affected by manure type, animal diet, the type of manure storage and handling, and manure application techniques. To help reduce GHG creation and work with large amounts of excess manure, it is important that manure management in the province concentrates on disposing manure in an environmentally and economically friendly manner.

Objectives for manure management should focus on maintaining or improving local water and air quality by limiting unpleasant odors, reducing nitrogen (N) and phosphorus (P) concentrations in manure and efficiently spreading manure. Although many management technologies exist, not all are realistic or cheap enough for farmers to implement. Different nutrient management strategies will work better for different farms.

## Manure Handling Systems

Emissions from manure handling systems are released when favourable conditions are met for gas creation. Warm, wet conditions tend to create higher amounts of both $CH_4$ and $N_2O$.

To reduce gaseous emissions when handling manure ensure that manure is not left in the barn environment for extended periods of time. Manure kept in a barn will tend to be warmer than manure stored outdoors and will produce more methane. Keeping barns clean and dry will help lower the loss of ammonia, reducing $N_2O$ production. Barn scraper systems can provide regular manure removal from the barn and store waste in proper storage areas.

Solid manure management systems, where poultry and livestock are housed on dry bedded manure packs of straw or sawdust were found to have lower $CH_4$ emissions when compared to liquid or slurry handling systems.

## Manure Storage Systems

Manure storage is one of the main areas where a farmer can control how many nutrients remain in or are lost from the manure. It is in the best interest of the farmer to focus on what method of storage system is best for the farm and type of livestock. Certain manure storage systems are more environmentally friendly than others, but may not be the best fit for the type of livestock the farmer is raising.

The following practices are encouraged:

- Avoid liquid or slurry handling systems – Methane production takes place when manure decomposes in the absence of oxygen (anaerobic conditions). Therefore, $CH_4$ production is higher in liquid manure.

- Use manure storage covers – Roofs for solid, covered tanks for liquid – to trap manure gases. In liquid systems, covers may reduce methane emissions by up to 95 percent. Covers also have the added benefit of controlling odour. Odour means that gas and money is being lost!!

- Avoid disturbing liquid manures in lagoons – Aerating lagoons increases oxygen and can eliminate $CH_4$ emissions, but increase $N_2O$ emissions.

- Avoid straw covers – Using a straw cover may be an effective odour barrier, but when the straw sinks into the liquid manure it adds C, which can substantially increase $CH_4$ production.

- Avoid stockpiling manure for long periods – Stockpiling can lead to anaerobic decomposition, resulting in both $CH_4$ and $N_2O$ emissions.

## Manure Testing

Manure testing should be done routinely to determine the amount of plant-available nutrients, particularly N and P.

Current legislation states that manure application be based on soil nutrient levels:

- When soil Olsen-P levels are between 60 and 180 ppm, manure can be applied no more than five times the annual crop removal rate of phosphate ($P_2O_5$).

- Additionally, nitrate-N levels can be no more than 140 lbs per acre (157.1 kg/ha) of soil class 1 to 3.

Because both nutrient levels are important in terms of the amount of applied manure, manure testing is a cost-effective farming practice.

## Proper Manure Application

Timing is everything when it comes to properly applying manure to fields. To ensure that crops receive the most nutrients possible, manure should be applied when crops will use it.

If manure is not taken up by plants, losses will occur through gaseous emissions, leaching or by surface run-off.

The following strategies can help improve manure nutrient use by crops and result in less GHG emissions.

- Apply manure to fields as soon as possible after removal from storage. Storing manure for long periods can encourage anaerobic decomposition and lead to increased $CH_4$ emissions.

- Inject or incorporate manure as soon as possible after application to reduce N loss.

- Avoid applying manure in areas where soil can become saturated, as this leads to anaerobic decomposition and increased $N_2O$ emissions.

- Eliminate winter applications to reduce the risk of run-off, and reduce the amount of nitrate-N in soils during spring snowmelt when risk of $N_2O$ loss is greatest.

- Spread manure uniformly around pasture to reduce N losses.

- Move winter feeding and bedding areas around pastures so manure is more evenly distributed. This will result in better decomposition.

- Station winter feeding areas on level ground away from riparian areas. This will reduce the risk of manure run-off entering surface watercourses.

## Eliminate Winter Spreading

Winter manure application should be eliminated to prevent manure runoff at spring-thaw and to reduce spring-thaw $N_2O$ emissions.

Effective November 10, 2013, the spreading of livestock manure between November 10 and April 10 in Manitoba is prohibited under The Environment Act: Livestock Manure and Mortalities Management Regulation, unless otherwise noted.

Applying manure after April 10 encourages farmers to apply manure at a time when their crops are just beginning to grow. The developing crop uses the nutrients as they become plant-available, minimizing the risk of loss to the environment.

Should moving manure during the winter be necessary, it is recommended that the manure be stock-piled in the fieldand spread following spring-melt.

## Composting Manure

Composting breaks manure into a more stable organic form, slowly releasing nutrients over time. Compost is rich in C, free from most pathogens and weed seeds, and improves soil nutrient status. Because compost reduces the amount of synthetic fertilizer needed on fields, composting manure helps lower net GHG emissions from livestock systems. The aerobic (with air) method of decomposing manure is also thought to lower $CH_4$ **and $N_2O$** creation. However, more research is needed to determine the exact benefit of composting manure net GHG reduction.

## Anaerobic Digesters

Anaerobic digestion is the oxygen-free process through which manure is broken down by microbes. The microbes produce a mix of $CH_4$ and $CO_2$, called biogas. This biogas can be cleaned and used as a natural gas replacement, burned as fuel or used by a generator to produce electricity or heat. The remaining organic material left after the digestion process has some nutritional value, very little odor and can be applied to fields as fertilizer.

Current anaerobic digesters on the market are much too expensive for most farmers to own. Research continues at the University of Manitoba to determine the benefits of digesters on manure management.

This digestion system works better for dairy and cattle manure, as poultry and swine manure presents more of a challenge due to their higher nitrogen levels. Anaerobic digestion is known to reduce pathogens, odour and weed seeds in the digested manure, reduce GHG emissions and provides an alternative fuel source.

Digesters may be the technology of the future to lower farm fuel consumption or provide alternative energy creation.

## Livestock and Manure Management

Poor manure management practices are common on much of the world's farms, as farmers lack awareness about the value of livestock manure as a fertilizer and fuel. Manure is often disposed of in piles, slurries or lagoons, which can lead to significant emissions of the greenhouse gas methane, as well environmental degradation, negative health impacts, and the loss of valuable nutrients that could be added to soil.

The Coalition's Agriculture Initiative promotes integrated manure and urine management practices that prevent harmful short-lived climate pollutant emissions while also providing added benefits for farmers through cost savings and additional income.

## Challenges

Livestock manure contributes to short-lived climate pollutant emissions through two processes:

- Through storage methods, especially liquid storage, which emit large amounts of methane.

- Through the burning of pastureland and the use of dung as a fuel for heating and cooking, which emit black carbon.

These pollutants in turn hinder agricultural production through their impacts on air quality and climate change.

The demand for livestock products, especially in developing countries, is expected to increase due to population growth and changes in dietary preferences. Without proper manure management practices, the increased animal numbers needed to meet this demand will result in an equal increase in emissions and other problems arising from manure collection, storage, treatment, and utilization.

While integrated manure management practices exist today, many farmers lack information to improve manure management or are faced with institutional, technical and socio-economic constraints that prevent them from adopting new practices.

Methane is the main greenhouse gas emitted by liquid manure in storage.

## Objectives

The Coalition aims to facilitate changes in manure management practices at the policy and local levels.

The Coalition supports integrated practices that encompass all activities associated with the management of dung and urine: excretion, collection, housing and storage, anaerobic digestion, treatment, transport, application, and losses and discharge at any stage along the 'manure chain'.

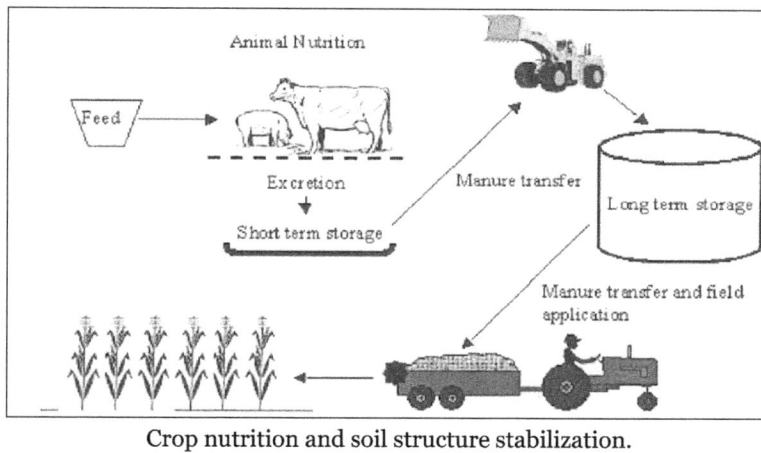

Crop nutrition and soil structure stabilization.

Key areas of work include:

- Potential reduction of methane emissions from mixed dairy farming systems.

- Beef production.

- The small ruminant sector.

- Commercial pig production.

- Improved mixed dairy systems.

Partners and stakeholders have coordinated efforts to reduce short-lived climate pollutant emissions by:

- Raising awareness of manure management options at the level of policy, private sector and farmers organizations through outreach and communication.

- Establishing an Advisory Board of leading international institutions to provide strategic guidance.

- Establishing a Central Hub and three Regional Centres – Africa, Latin America, and Asia - working in close collaboration, to identify opportunities and conduct work in regions, build networks and partnerships, gather information, and implement projects.

- Establishing networks to exchange manure management information, connect people, and forge partnerships.

- Launching an on-line knowledge hub - Manure Management Kiosk – to provide remote access to best practices about manure management.

- Establishing a roster of experts to provide targeted technical assistance and training, analysis and practical implementation and policy support, relying heavily on co-financing and in-kind resources from partners.

- Launching projects and partnerships to improve manure management by providing information, experts, knowledge exchange, and access to resources.

Key objectives include:

- Integrating manure management practices into livestock systems.

- Improving existing practices to reduce short-lived climate pollutants and other harmful emissions to the environment.

- Capturing methane as an energy source.

- Optimizing nutrient utilization for crop production by managing and removing barriers to action with a view toward enhancing food security and sustainable development.

## Impacts and Results

Coalition-funded work to date has resulted in:

- 7 Opportunities for Practices Change (OPCs) in seven countries that include examples of actual changes in manure policies and manure management practices.

- 13 trainings for extension workers, with almost 300 extension workers trained.

- 5794 farmers directly trained by the above extension workers on better manure management to reduce SLCPs.

- 2 policies in integrated manure management drafted, 1 policy in development; 1 policy influenced.

- 52 regional events and 15 global events or meetings to raise awareness, with an estimated 2000 stakeholders directly reached on manure management.

Together these activities have helped policymakers, practitioners and other key stakeholders acquire, share, and disseminate knowledge and leverage new and existing resource bases and institutions to enable the adoption of improved manure management practices at the regional, national, and local level.

## Livestock Disease Management

Livestock systems in developing countries are characterised by rapid change, driven by factors such as population growth, increases in the demand for livestock products as incomes rise, and urbanisation. Climate change is adding to the considerable development challenges posed by these drivers of change. The increasing frequency of heat stress, drought and flooding events could translate into the increased spread of existing vector-borne diseases and macro-parasites, along with the emergence of new diseases and transmission models. Appropriate sustainable livestock management practices are required so that livestock keepers can take advantage of the increasing demand for livestock products (where this is feasible) and protect their livestock assets in the face of changing and increasingly variable climates.

Livestock diseases contribute to an important set of problems within livestock production systems. These include animal welfare, productivity losses, uncertain food security, loss of income and negative impacts on human health. Livestock disease management can reduce disease through improved animal husbandry practices. These include: controlled breeding, controlling entry to farm lots, and quarantining sick animals and through developing and improving antibiotics, vaccines and diagnostic tools, evaluation of ethno-therapeutic options, and vector control techniques.

Livestock disease management is made up of two key components:

- Prevention (biosecurity) measures in susceptible herds.

- Control measures taken once infection occurs.

The probability of infection from a given disease depends on existing farm practices (prevention) as well as the prevalence rate in host populations in the relevant area. As the prevalence in the area increases, the probability of infection increases.

## Prevention Measures

Preventing diseases entering and spreading in livestock populations is the most efficient and cost-effective way of managing disease. While many approaches to management are disease specific, improved regulation of movements of livestock can provide broader protection. A standard disease prevention programme that can apply in all contexts does not exist. But there are some basic principles that should always be observed. The following practices aid in disease prevention:

- Elaboration of an animal health programme.

- Select a well-known, reliable source from which to purchase animals, one that can supply healthy stock, inherently vigorous and developed for a specific purpose. New animals should be monitored for disease before being introduced into the main flock.

- Good hygiene including clean water and feed supplies.

- Precise vaccination schedule for each herd or flock.

- Observe animals frequently for signs of disease, and if a disease problem develops, obtain an early, reliable diagnosis and apply the best treatment, control, and eradication measures for that specific disease.

- Dispose of all dead animals by burning, deep burying, or disposal pit.

- Maintain good records relative to flock or herd health. These should include vaccination history, disease problems and medication.

## Surveillance and Control Measures

Disease surveillance allows the identification of new infections and changes to existing ones. This involves disease reporting and specimen submission by livestock owners, village veterinary staff, district and provincial veterinary officers. The method used to combat a disease outbreak depends on the severity of the outbreak. In the event of a disease outbreak the precise location of all livestock is essential for effective measures to control and eradicate contagious viruses. Restrictions on animal movements may be required as well as quarantine and, in extreme cases, slaughter.

Holistic approaches to disease prevention control (woman and man participants in rural training course in learning how to improve health of their goats - Sudan).

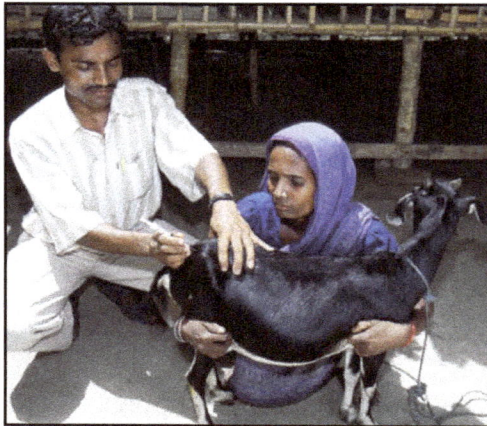

Holistic approaches to disease prevention control
(Man immunising goat held by woman - Bangladesh).

The major impacts of climate change on livestock diseases have been on diseases that are vector-borne. Increasing temperatures have supported the expansion of vector populations into cooler areas. Such cooler areas can be either higher altitude systems (for example, livestock tick-borne diseases) or more temperate zones (for example, the outbreak of bluetongue disease in northern Europe). Changes in rainfall pattern can also influence an expansion of vectors during wetter years and can lead to large outbreaks. Climate changes could also influence disease distribution indirectly through changes in the distribution of livestock. Improving livestock disease control is therefore an effective technology for climate change adaptation.

## Advantages of the Technology

Benefits of livestock disease prevention and control include: higher production (as morbidity is lowered and mortality or early culling is reduced), and avoided future control costs. When farmers mitigate disease through prevention or control, they benefit not just themselves but any others at risk of adverse outcomes from the presence of disease on that operation. At-risk populations include residents, visitors and consumers. The beneficiaries might also include at-risk wildlife populations surrounding the farm that may have direct or indirect contact with livestock or livestock-related material.

## Disadvantages of the Technology

Management options may interact, so the use of one option may diminish the effectiveness of another. Another critical issue is the long-term sustainability of currently used strategies. Chemical intervention strategies such as antibiotics or vaccines are not biologically sustainable. Animals develop resistance to drugs used to control certain viruses and with each new generation of vaccine a new and more virulent strain of the virus can arise. Small-scale producers may be negatively affected by livestock disease management if the full cost of the disease management programme is directly passed onto them with no subsidy from the government.

## Financial Requirements and Costs

Livestock disease management costs include: testing and screening, veterinary services, vaccines, training of livestock keepers and veterinary staff, and perhaps changes to practices and facilities to reflect movement restrictions and quarantines when animals are added to the herd.

## Control of Mastitis

A low-cost technology applicable to a wide range of livestock (cattle, sheep and goats) is the control of mastitis. Mastitis is an infectious disease caused by pathogenic micro-organisms due to inadequate milking practices or blows to the udders. It is one of the diseases that cause the most financial losses in milk production. In conditions of increasing climate variability, emergence of new pests and diseases can introduce invasive organisms to the livestock environment. It is therefore essential that livestock farmers are able to identify and prevent mastitis in order to maintain healthy animals that, in turn, are more capable of withstanding adverse weather conditions such as prolonged droughts or severe frosts.

Information and monitoring requirements for the control of mastitis include:

- Producer training on testing and diagnosing mastitis, hygienic milking practices, teat sealing, treatment of clinical mastitis, control records.

- Organisations or institutions must have extension farmers or technicians who are trained in the mastitis control process.

- Monitoring and regular check-ups are necessary for the prevention of mastitis.

The following is also required in the application of this technology:

- The California Mastitis Test (CMT) or black background rate. This is very easy for farmers to use as readings are immediate and low cost.

- Teat sealant to protect the udder against mastitis germs.

- Clean and disinfected containers, cloths and mechanical milking machines.

- Milking records which should contain basic information like the name of the animal, the date, the name of the person milking the animal, the rooms examined, evidence of mastitis, density and acidity of the milk.

Institutional and organisational requirements must also be taken into account: health care institutions and producers' organisations should carry out sanitation campaigns, hold training workshops and provide technical assistance on the control of mastitis, using adequate informative materials like easy-to-read leaflets and flyers that the cattle farmers can understand and follow. Costs and financial requirements are relatively low. The CMT costs about US$25 and can last about six months for an average of three cows per farmer. The teat sealant costs about US$30.

Prevention and control costs are generally evaluated against expected financial losses resulting from a disease outbreak in a cost-benefit analysis. The assumption is that increased prevention and control costs lower the expected losses by diminishing the expected scale of an infection. McInerney et al present the problem graphically as a cost minimisation problem:

$$\min C = L + E$$

Where C is total annual disease cost, L is the value of output losses, and E is the control expenditures (which themselves are a function of inputs purchased for control).

## Institutional and Organisational Requirements

Countries should cooperate in programmes against trans-boundary disease either through formally formed organisations or networks. Neighbouring countries often have similar production systems and disease risk profiles and will be more likely to be affected by similar climate change impacts in livestock disease. There will be mutual benefits and cost savings through joint preparedness planning. Public policies range from bounties/indemnities for infected livestock to required herd depopulation and farm decontamination, to decentralisation programmes for provision of veterinary services and drug supplies. Livestock and animal health policy should be oriented to both the commercial and pastoral sectors and include pro-poor interventions to support the most vulnerable populations. Government investments in infrastructure (including early warning systems, roads, abattoirs, holding pens, processing plants, air freight/ports and so on), systematic vaccination, and in research and development can all contribute to providing an enabling environment for effective livestock disease management. Removing or introducing subsidies for improved management, insurance systems and supporting income diversification practices could benefit adaptation efforts.

In order for producers to make decisions regarding disease management, they must understand the options that they have. These options depend on disease biology, prevention techniques, tests for infection and their costs, treatments available, market reactions, as well as industry and government programmes and policies. Disease biology includes transmission modes and rates, disease evolution (for example, length of time to infectious period), production losses associated with the disease, and mortality rate (where applicable).

Practical training for farmers should include:

- Principles of anatomy and physiology of the livestock animals.

- Principles of nutrition and pasture ecology.

- Animal diseases of local importance: clinical and post mortem signs, epidemiology, prevention, treatment. Applying first aid, the use of basic veterinary medicines (wound treatments, dips, anthelmintics, antibiotics, trypanocides, babesiacides, vaccines, care and storage of medicines and vaccines, and the use and care of syringes).

- The basic principles of sero-surveillance campaigns — how to draw blood and store sera.

Modelling disease outbreaks and spread can provide valuable information for the development of management strategies. Modelling involves studying disease distribution and patterns of spread to determine the scale of a problem. This information is used to develop a model that can predict the spread of disease. Disease modelling requires prior knowledge of animal population distributions and ecology, diseases present and methods of disease transmission. Modelling can be used to assess potential disease impacts and develop contingency plans.

Geographic Information System (GIS) software can play a key role in livestock disease management. The main advantage of GIS software is not just that the user can see how a disease is distributed geographically, but also that an animal disease can be viewed against other information. For example, maps that show possible impacts of climate change on rainfall patterns, crop yields and flooding. The disease presence can then be related to these factors and more easily appreciated visually. This is important in relation to managing and responding to the changes in distribution of diseases due to changing climate.

## Role of Indigenous Knowledge in Livestock Disease Management under Climate Change

Indigenous knowledge about livestock disease management has been shown, in certain cases, to be cost-effective, sustainable, environmentally friendly and practical. Practices include:

- Utilisation of local plant remedies for prevention and cure of diseases.

- Avoiding certain pastures at particular times of the year; and not staying too long in one place to avoid parasite build-up.

- Lighting smoke fires to repel insects, especially tsetse flies.

- Mixing species in the herd to avoid the spread of disease.

- Avoiding infected areas or moving upwind of them; spreading livestock among different herds to minimise risks; and quarantining sick animals.

- Selective breeding. As an example from the arid south of Zambia, restocking and promoting the rearing of drought-tolerant goat breeds are adaptive measures already being undertaken.

## Barriers to Implementation

A lack of strong institutions and political will to monitor disease status effectively can produce a considerable barrier to livestock disease management. Difficulties in eradication of disease may also be exacerbated by many small-scale and backyard producers, infected wildlife, smuggling, and cockfighting. If there is no compensation for stamping out disease through slaughter, then producers, particularly small-scale producers, may be reluctant to participate. If they do participate it may mean that they no longer can afford to produce.

## Opportunities for Implementation

Where the disease organism has built up resistance against vaccines or the animal has built resistance against the disease there is an opportunity for incorporating simple, high-tech genetic approaches such as selective breeding. National planning for livestock disease management also presents an opportunity to improve agricultural support services in rural areas and to incorporate indigenous knowledge into formal prevention and control plans, thereby unlocking the potential of low-cost interventions and disseminating information on traditional lessons and experiences to a wider audience. Trans-border collaboration can provide an opportunity to strengthen veterinary services and can improve the effectiveness of disease management programmes through harmonisation of prevention and control measures, such as disease reporting and surveillance.

# Breeding Management of Cattle and Buffaloes

Reproduction is an important consideration in the economics of cattle production. In the absence of regular breeding and calving at the appropriate time, cattle rearing will not be profitable. A healthy calf each year is the usual goal. This is possible only by increasing the reproductive efficiency of the animals.

Successful reproduction encompasses the ability to mate, the capacity to conceive and to nourish the embryo and deliver the viable young ones at the end of a normal gestation period. In fact, interruption in this chain of events leads to failure of the cow either to conceive or the embryo to die or to have a premature delivery of the foetus.

The reproductive efficiency is a complex phenomenon controlled by both genetic and non-genetic factors, the non- genetic factors being climate, nutrition, and level of management. The reproductive efficiency varies not only between species and breeds but also among the animals within the same breed. Even the best feeding and management can not coax performance beyond the genetic limit of an inferior animal. Improving the genetic merits of livestock populations is important at all levels of management. A sound breeding programme is a necessary part of the total animal production system.

### Breeding Efficiency

### Factors affecting Breeding Efficiency

The factors which influence the breeding efficiency of cattle are as follows:

## Number of Ova

The first limitation on the breeding efficiency of fertility of an animal is the number of functional ova released during each cycle of ovulation. Ovulation is the process of shedding of ovum from the

Graffian follicle. In the case of cow, usually a single ovum is capable of undergoing fertilization only for a period of 5-10 hours. Therefore, the time of mating insemination in relation to ovulation is important for effective fertilization.

## Percentage of Fertilization

The second limitation is fertilization of ova. Failure to be fertilized may result from several causes. The spermatozoa may be few or low in vitality. The service may be either too early or too late. so that the sperms and eggs do not meet at the right moment, to result in fertilization.

## Embryonic Death

From the time of fertilization till birth, embryonic mortality may occur due to a variety of reasons. Hormone deficiency or imbalance may cause failure of implantation of fertilized ova which die subsequently. Death may occur as a result of lethal genes for which the embryos are homozygous. Other causes may be accidents in development, over-crowding in the uterus, insufficient nutrition or infections in tile uterus.

## Age of First Pregnancy

Breeding efficiency may be lowered seriously by increasing the age of first breeding. Females bred at a lower age are likely to appear stunted during the first lactation, but their mature size is affected little by their having been bred early.

## Frequency of Pregnancy

The breeding efficiency can be greatly enhanced by lowering the interval between successive pregnancies. The wise general policy is to breed for the first time at an early age and to rebreed at almost the earliest opportunity after each pregnancy. In this way the lifetime efficiency is increased. Cows can be rebred in 9-12 weeks after parturition.

## Longevity

The length of life of the parent is an important part of breeding efficiency, because the return over feed cost is greater in increased length of life. Also, it affects the possibility of improving the breed. The longer the life of the parents, the smaller the percentage of cows needed for replacement every year.

## Management Practices to Improve Breeding Efficiency

Some of the management suggestions which will tend to improve breeding efficiency of cattle are listed below:

- Keep accurate breeding records of dates of heat, service and parturition. Use records in predicting the dates of heat and observe the females carefully for heat.
- Breed cows during near the end of mid heat or heat period.
- Have females with abnormal discharges examined and treated by veterinarian.

- Call a veterinarian to examine females not settled after three services.

- Get the females checked for pregnancy at 45 days to 60 days after breeding.

- Buy replacements only from healthy herds and test them before putting them in your herd.

- Have the females give birth in isolation, preferably in a parturition room and clean up and sterilize the area once parturition is over.

- Follow a programme of disease prevention, test and vaccination for diseases affecting reproduction and vaccinate the animals against such diseases.

- Practice a general sanitation programme.

- Supply adequate nutrition.

- Employ the correct technique.

- Provide suitable shelter management.

- Detect silent or weak heat, by using a teaser bull.

## Selection and Culling

Selection and culling are the two sides of the same coin. Selection is the process in which certain individuals in a population are included for becoming the parents of the next generation. Automatically some are excluded for the purpose which are culled. Natural selection has been going on since ages where animals which were stronger, which had better survivability and which were in more unison with the environment around them, found a better chance to reproduce.

Thus certain genes for certain characters got more chance to be selected to form individuals in the subsequent generations. Since domestication of cattle, man has been looking for superior phenotypes in traits useful to him and selecting such animals to form the parental generation. This is man made artificial selection. Now man has progressed one step further in making estimates of genotypes from the study of phenotypes and making use of that information (in artificial) selection.

## Selection Methods

There is only one way to select and that is to "keep the best and cull the poorest. The various selection methods are techniques for identifying or estimating the genetic values of individual candidates for selection. The procedure discussed here apply to selection for quantitative trails.

## Performance Testing

Performance test is a measure of the phenotypic value of the individual candidates for selection. Since the phenotypic value is determined by both genetic and environmental influences, the performance test is an estimate, not a measure of the genetic value. The occurrence of this estimate depends upon the heritability of the trait i.e. on the degree to which the genetic value is modified by the environmental influences.

## Advantages

- Among simple procedures, the performance test is the most accurate.

- Environmental influences can be minimised by testing candidates for selection in the same pen or in similar environmental conditions.

- The measure is direct, not on a relative basis.

- All candidates for selection can be tested in contrast to progeny testing where only a parent can be tested.

- Generation intervals are usually short.

- Testing can usually be done on the farm under normal management conditions.

## Disadvantages

- Accuracy become low when heretability is low.

- Phenotypes are not available for one sexor in sex limited traits such as milk yield.

- Traits which are not expressed until maturity may become expensive or difficult to manage by performance tests since most selection decisions must be made before maturity.

- Performance tests should be the backbone of most selection programmes. Although much publicity has been given to other selection methods, it remains a fact that most of the progress in livestock improvement to date has been due to selection on the individual's own phenotype i.e. performance test.

## Pedegree Selection

A pedegree is a record of an individual's ancestors including its parents. This information is valuable because each individual possesses a sample half of the genes from each parent. If we can precisely know an individual's phenotype, little is gained by considering pedegree in selection. Pedegree considerations are useful when we do not have sufficient accurate records of production of the individual. Also, it is useful in the early selection when the traits in question might not have expressed themselves. It is also useful for selection of males when the traits selected for are expressed only by the female such as milk production in dairy cattle.

## Advantages

- It provides information when performance tests are not available for the candidates.

- It provides information to supplement performance test information.

- It allows selection to be completed at a young age. Pedegree records may be used to select animals for performance or progeny testing in multi-stage selection scheme.

- It allows selection of bulls can be selected on the milk records of their female relatives.

## Disadvantages

- Accuracy, relative to alternative selection procedures is usually low.

- Too much emphasis on relatives, especially remote relatives, greatly reduces genetic progress.

- Progeny of favoured parents are often environmentally favoured.

- Relatives often make records under quite different environments, thus introducing non random bases into the selection system.

## Progeny Testing

In this method we evaluate the breeding value by a study of the expression of the trait in its offsprings. Individuality tells us what an animal seems to be, his pedigree tells us what he ought to be, but the performance of his progeny tells us what he is.

Progency testing is, of course, a two-stage selection system because some preliminary selection determines which animals first produce progeny followed by further culling of these which produce poor progeny.

## Advantages of Progney Testing

- High accuracy when many progeny are obtained.

## Disadvantages Progney Testing

- Long generation interval.

- Requires high reproductive rate.

- Low selection intensity.

## Show Ring Selection

Selection on the basis of show ring performance has had considerable value in the past. Essentially this selection has been directed towards bringing the conformation of the animal to some ideal conformation.

This improvement has been based on two goals:

- Improvement conformation, and

- Correlated response.

Improvement of conformation has economic value because a part of the sale price is determined by the conformation of the individual. The ideal type was chosen so that, in the opinion of the judges, the animal possessing this conformation was most likely to be a profitable producer. In other words, the judges were attempting to stress traits of conformation which are corrected with productive ability.

With the advent of record keeping it was found that direct selection for performance traits resulted in much faster progress than selection through correlated conformation traits. Also, when subjected to intensive study, many of the correlations between performance and show ring were found to be of non-genetic origin.

If the correlations are of genetic origin, direct selection for performance should improve conformation as well as the reverse situation. The show ring has been a good forum for discussion of what constitutes ideal type and good management and has produced dramatic changes in the conformation of some species.

This has resulted primarily from education of the breeders, however, for most animals which are presented in the ring are good and selection differential among these animals is usually so small as to produce little change.

## Advantages of Show Ring Selection

- It enables breeders to exchange ideas and experience.
- It allows comparisons among superior animals both within and between breeds.
- It allows new breeders to make contact with established breeders.

## Disadvantages of Show Ring Selection

- Emphasis is usually placed on traits of little economic importance.
- Clever fitting and showmanship can mask defects of various kinds.
- Differences between exhibited animals are usually small.
- Conformation and production traits usually have low genetic correlations.

## Choosing Traits for Selection

Many factors enter into the choice of traits to be selected for. The following ones are the most important.

- The goal of the selection programme.
- The habitability of the traits.
- The economic value of improvement in each trait.
- The range in variation of each trait.
- Correlation among the traits.
- The cost of the selection programme.

## Selection Goals

Often the goal of the selection programme makes the choice of traits quite obvious. The breeder of the race horses must select for speed if he is to be successful and his choice of traits are limited to alternative ways to measure speed. Similarly, the breeder of dairy cattle generally sets out to breed cows with superior milk production characteristics. Thus, his choice of traits is specified by his selection goals.

## Heritability

Heritability is defined to be the fraction of the superiority of parents which is, on the average, transmitted to their off-springs. To explain habitability in simpler words: Heritability tells us how much of the observable differences in the animal is caused by genes and how much by environment.

Heritability for the same characteristics may vary from one population to another and also may vary from one characteristic to another even ink the same population. The ability to recognise the breeding values or transmitting abilities of animals is closely associated with heritability. If the heritability is high for a trait, we can proceed straight way to adopt a system of mass selection of superior animals, with little attention to pedegree information, collateral relatives, progeny test or inbreeding and genetic improvement in that trait is low, genetic progress may be disappointing with mass selection and greater attention should be paid to pedegree records, family information and use of progeny tests.

## Variability of the Trait

Selection operates on the variability in expression of the trait uniform for a trait. there will be little selection response because any selected groups of parents will not be much better than those not selected. Some traits are much more valuable than others. thus the innate variation of the traits should be carefully considered in choosing traits for selection.

Variation can be increased by improving exotic types and sometimes this can result in new combination of genes which are superior to either parent type.

## Correlated Traits

Sometimes traits tend to be inherited together. These correlations may arise in several ways.

The traits may be of different measures of some underlying trait. For example. weight and height are both measures of body size. thus taller animals are usually heavier and these two traits are said to be correlated.

If the same genes produce response in several traits. those traits will be correlated. This condition is referred to as pleiotrophy.

Correlated responses are common. Selection for increased milk yield produces a correlated decrease in the per cent of fat in the milk of dairy cows. Thus. both direct and correlated responses result from selection and some correlated responses are positive while others negative.

Correlated response may be advantageously used in selection programme. For example feed efficiency is expensive to measure because it requires both weight gain and feed intake on each individual, whereas weight gain requires neither feed weight nor individual feeding.

The definite goals are essential for a successful selection programme. The success in achieving these goals depends on the existence of genetic differences. the degree to which phenotype differences are heritable and the correlated responses in other traits. In comparing the selection programme, the breeder must evaluate the value of the expected response and the cost of the programme relative to the costs and responses of alternative selection programmes.

## Systems of Breeding

The ultimate aim of the breeder is to evolve outstanding and improved type of animals which can render better service to man. Selection and system of breeding constitute the only tools available to the breeder for improvement of animals. Since new genes can not be created though they can be recombined into more desirable groupings.

Systems of breeding has been broadly divided as under:

- On breeding - Breeding of the related animals.

- Out breeding - Breeding of the unrelated animals.

## lnbreeding

Inbreeding is a mating system in which individuals mated are more closely related than the average of the population from which they come. It means the mating of males and females which are related. Animals deemed to be related only when they have one or more ancestors in common on the first 4-6 generations of their pedegree. The intensity of inbreeding depends upon the degree of relationship. Close inbreeding denotes mating of closely related individuals like dam to son (mother x son) or sire to daughter (father x daughter) or full brothers to full sisters.

In breeding makes more pairs of genes in the population homozygous. Wherever there is inbreeding, there will be one or more common ancestors from which, part of the gene samples (gametes) have arisen.

Inbreeding can again be divided into following groups:

## Estrus Period

- Pro estrus: 2 or 3 days.

- Estrus: 12 to 18 hours.

- Ovulation: 12 to 16 hours after the end of estrus.

- Estrous cycle length: $21\pm 3$ days.

## Puberty

- Puberty is the stage at which animal becomes sexually mature and secondary sex characteristics become conspicuous.

- The term sexual maturity means that the animal is capable of reproduction.

- Puberty is the age at which the first estrus occurs in the heifer and the bull starts giving semen with viable sperms.

- The reproductive organs undergo marked increase in size at the time of puberty.

- Under good feeding a calf attains puberty approximately at 66 per cent of adult body size.

## Signs of Estrus

- Cow in estrus will be the first cow to rise in the morning.

- The cow become restless does not eat and frequently bellows and seldom ruminates.

- Sudden drop in milk production.

- Searching for male.

- Traits of homosexuality is shown in which the cow will attempt to mount other cows while other females not in estrus tend to mount the estrus cow which she permits.

- The cow is receptive to the act of mating and will stand when the bull mounts her.

- The behavior of standing quietly while being mounted by the bull or other cow is referred to as the 'standing heat' which is the surest sign of estrus.

- This extends for 14-16 hours and shows other symptom like bellowing, nervousness, anorexia, reduction in milk yield.

- Mucous discharge may be found sticking to the tail.

- In early heat the discharge is watery and copious in mid heat (standing heat) it becomes thick and sticky and in late heat it will be scanty and discoloured.

## Bulling

The best indicator of oestrus is when any cow or heifer repeatedly stands and accepts mounting by one of her herd mates. Unfortunately, they do not do this on demand. Those responsible for oestrus detection must watch for this behavior and combining what they see with their own previous knowledge/experience, to decide whether to inseminate or not.

## Heat Detection in Buffaloes

- Cows do mount over other cows when they are likely to come in heat and stand for mounting when they are in good heat. This is not seen in buffaloes. Buffaloes neither mount on other buffaloes nor other buffaloes mount on buffaloes in heat.In buffaloes copious ropy hanging discharge is not seen on the contrary it gets suddenly dropped and is not noticed by the owner and the discharge is scanty.Some buffaloes do not bellow and show silent heat, especially high yielding buffaloes.

- The main heat symptoms of buffaloes are as follows.The vulva becomes edematous, swollen. The lower portion of vulva looks oily. The gap is seen between vulvar lips and slight opening is seen. The wrinkles which are present in anoestrus buffalo become shallow or vanish.

- The mucous membrane of vulva becomes reddish, moist and glossy.

- Mucus discharge which is not seen normally can be seen before or after oestrus spontaneously.

- The colour, consistency and fern pattern of mucus help in determination of correct oestrus.

- Engorgement of teats in lactating buffaloes which is due to holding of milk following increased estrogen level in blood is seen when they are in heat.

- Frequent urination. The urine coming in spurting action wetting the part of skin below vulva and above udder (perineum). The drying of the urine leaves white mark on skin.

- Buffaloes in heat remain restless, off feed, raising head in a typical fashion.

- Local non descript buffaloes bellow, become restless and remain off feed. Milk yield is reduced. The bellow is sharp and for longer duration.

- The buffaloes expose their teeth while bellowing which is very characteristic.

- The mucus discharge, in buffaloes is seen in about 49% cases. It is thin on the day of heat, become thick as the time passes and changes the colour from clear to white.

- 60-70 % of the buffalo come in heat from 6 pm to 6 am (after sunset and before sunrise) and this should be borne in mind and attendant should watch the buffaloes in the evening and early morning for expression of heat symptom.

- Teaser bull (Vasectomised bull), can be used for parading in buffalo barn for detection of heat.

## Other Methods to Detect Estrus

- Crystallization pattern of cervical mucous will show long crystals in a typical fern-like pattern.

- There are many estrus-detecting devices available. They are usually attached to the tail or rump of the cow.

- Mounting causes these devices to discharge a coloured fluid which can be observed afterwards even from the distance.

- 'Chin ball mating device' can be used for heat detection. It works on the same principles of a ball point pen and is fixed by means of a halter below the chain of the teaser bull. When the checking animal mount the cow in heat, the dye exuded round a spring-loaded ball of the device marking the back of the cow.

- Russian workers have developed an instrument basically consisting of an ohm meter and electrodes. When applied to the mucous membrane of the vagina, the resistance indicated on the ohm meter shows whether the cow is in heat.

- Pedometer is an instrument used to monitor the movement of animal. The principle is the activity and movement of the cow increases on the day of heat and this can be detected by means of a pedometer tide to the leg of the cow.

- The vaginal temperature can be recorded, which gives an indication about the heat. Generally during estrus, the vaginal temperature increased by about 1°Con the day of heat. Both methods are not very practicable.

- The methods described above had little applicability in developing countries due to technological and economical and managemental reasons. close observation of signs of heat, standing heat remains the most practicable method of heat detection.

- In large farm this can be supplemented with a bull-parade using a teaser bull. A teaser may be a vasectomised bull or bull in which penis has been amputated and the urethra exteriorized.

- An intact bull can also be used by hanging a thick cloth or gunny bag curtain in front of the penis preventing entry of penis and mating.

- Special care should be taken to prevent spread of disease by teasers. Vasectomised bull is more harmful in this regard.

## Time of Insemination

- Ovulation takes place about 12 hours after the end of estrus. It takes another six hours for it to travel half-way down the oviduct.

- The sperm, even though reach the oviduct within minutes after insemination, must be exposed to the female reproductive tract for about 6 hours to attain the capacity to fertilize.

- This process of preparing the sperm for fertilization is known as capacitation.

- Sperm are viable for 24 hours in the female reproductive tract whereas the ovum remains fertile for only about 10 hours after ovulation.

- This implies that mating or insemination between mid-estrus (middle of standing heat followed by another insemination in about 6 hours after that.

- As a routine practice, if a cow is seen showing signs of early heat in the morning, it may be inseminated in the evening.

- If such signs are first manifested in the evening, the cow may be bred next day morning.

- A cow is expected to show estrus in 30-40 days after calving. Cows that fail to show heat even after 50 days have generally some problem and need examination.

- It may be due to infection or malnutrition and remedial measures may be taken accordingly.

- Insemination should be done only when buffalo is in standing heat. In buffalo to understand standing heat one should know the symptoms of heat.

- Buffaloes normally are not seen standing for mounting by herd mates but standing heat can be known from the changing colour of mucus discharge which is early estrus is clear and watery but in standing heat or mid heat the colour is changed to little buffy with thick consistency.

- In mid heat the oedema of vulva is intense there is little gap in vulvar lips and lower lip looks oily.

- The vulvar mucus membrane is glossy reddish or pink and wet.

## Signs of Approaching Parturition

- Cow will leave the herd and seek isolation.

- Loss of appetite and distress.

- Distention of teat and udder, considerable milk appears in the udder and there may be dripping of milk.

- Relaxation of pelvic ligament one day before calving, the ligament on the sides of the tail head is loosened so that hollows appear on either side of the backbone and the tail head is raised and the quarters are dropped.

- The vulva become enlarged and flabby.

- Animal will be restless and will pace about often trying to kick or scratch the flank region.

- The parturition process has three stages a. preparatory stage (uterine contraction and dilatation of cervix) b. active expulsive stage c. expulsion of foetal membrane.

- Cow will deliver the calf within 12 hours after commencement of first stage and lapse in this vaginal examination of assistance by a veterinarian is required.

- Care must be taken to observe expulsion of placenta (after birth). It should be removed immediately so as to avoid cow eating it.

## Package of Practice to Improve Reproductive Efficiency

- Accurate record kept is very important in ensuring reproductive efficiency the herd.

- There production detail like date of estrus, date of service and calving should be maintained properly.

- This data should be used to predict the probable date of heat, such animal should be watched carefully in the morning and evening for signs of heat.

- In larger dairy farm teaser bulls can be put in use.

- Complete breeding history, past performance and difficulties of a individual cow should be maintained.

- Irregular estrus and abnormal discharge should be attended immediately.

- The cows with retained placenta should be treated promptly and when such cows are put in breeding next time, the reproductive tract should be examined thoroughly for involution and possibility of infection.

- A manager should examine a cow 24 to 36 hours after service for metestrus bleeding. If it occurs under 24 hours after service, the cows were bred too late.

- If it occurs over 36 hours after service, they were bred too early during estrus. This will help in pinpointing of failure of conception.

- Cow should be examined for pregnancy 45 to 60 days after service so that if they are non-pregnant, steps can be taken to re-breed them at the earliest opportunity.

- If the conception rate under A.I is lower than natural service, time of insemination, insemination technique and quality of semen must be checked.

- Short irregular cycles indicate cystic ovaries, short and long irregular interval point to missed head.

- Silent or quiescent heat : The behavioral manifestation of heat may be very weak or imperceptible in such case. It is very common in buffaloes. But there is a normal ovulation and if inseminated at proper time the animal can conceive.

- Cows go through the normal ovarian changes of the estrus cycle except the behavioral heat and sexual receptivity.

- It is more in summer season than other seasons and more in heifer than adult animals.

- Use of balanced feed, proper summer management, use of teaser bulls can be of use tool in detecting silent heat.

- Anestrus or absence of sexual cycle may be due to under developed genitalia or due to persistent CL. In the former case follicle fail to develop and a heifer will not come to heat at all.

- One of the major causes of under developed genitalia is malnutrition. Besides there can be genetic causes.

- The second probability is the anestrous also may be due to persistent CL, due to certain hormonal disturbances, the C L persist beyond the life expectancy in a normal cycle, thereby preventing further cycling. A common cause for persistent CL is endometritis of the uterus.

- Sometime anestrus is often observed in the early post partum period when the lactation is strong, probably due to the influence of lactation(due to secretion of prolactin) 'e'lactational anoustnum.

## Care and Management of Young Stock

- Normally newborn animals will be taken care by its mother and required little assistance.

- In case of cattle, sheep and goat immediately after birth the mucus around the nostrils should be whipped out using dry cloth or a hand full of straw can be used for this purpose.

- Calm environment should be provided to the mother and young animals for development of bond.

- The mother should be allowed to lick the newborn; if the dam fails to lick it can be stimulated by sprinkling small quantity of salt or bran over the young one.

- Immediately after birth the naval cord should be ligated with clean sterile cotton thread 1 inch from the body and tincture iodine should be applied to the naval cord.

- With in 1 hour after birth the newborn will able to stand and it should be allowed to drink adequate quantity of colostrum (first milk) which will give immunity to the newborn.

- Young animals should be housed comfortably. Adequate care should be taken to avoid housing young stock with adult stocks.

- In winter condition adequate warmth condition should be provided.

- Adequate bedding materials like straw or hay should be provided to newborn animals.

- For giving extra heat artificial light source can be utilized.

- Proper light, ventilation and hygiene should be maintained to avoid spread of disease.

## Care and Management of Dry Animals

- A dry animal means animal, which completes their lactation and drying is essential to give adequate rest to the udder of the animal.

- Dry animals should be separated from other milch animals.

- In case of cow, dry cow can be treated for mastitis to prevent mastitis in next lactation.

## Care and Management of Pregnant Animals

- Pregnant animals should be provided with extra ration to meet the requirement of fast growing foetus as well as store energy for future lactation.

- Pregnant animals should be separated in advanced stage from other non pregnant animals.

- They should be housed separately in place called calving pen.

- Adequate bedding materials should be provided in the pregnant animals shed.

- Floor of the pregnant animal shed should be non-slippery.

- Adequate clean fresh drinking water should also be provided in the calving pen.

- In advance stage of pregnancy laxative diet should be provided.

## Care and Management of Bullock

- Bullocks are normally used for agricultural operations and or transport purpose.

- Some bullocks are ferocious and so control them properly with nose rope or nose rings.

- The hooves of the bullocks should be provided with metal shoes to protect the hooves from wear and tear.

The working hours for bullocks are recommended as follows:

- Normal Work - 6 hours of carting or 4 hours of ploughing.

- Heavy Work - 8 hours of carting or 6 hours of ploughing.

- Sufficient roughages and 1-2 kgs of concentrates may be provided for feeding of bullocks during break period in works, the animal may be left for free grazing.

- The bullocks are housed in separate sheds with sufficient space and protection from hot and cool conditions.

- Free access to drinking water is essential. Regular grooming of animals should be practiced.

## Management of Bulls

- The bull is half of the herd. Not only the bulls should be genetically superior quality, but they also have to be in prime breeding condition by proper feeding and management.

- Bulls should be selected based on their pedigree and the bull calves should be separated from breedable cows and heifer by the time of attainment of puberty, which is between 1 ½ to 2 ½ years in zebu and buffalo breeds and still lower in crossbreds.

- The bull calf should be dehorned within a few days of birth by disbudding with chemical or hot iron.

- This practice is considered to make the bull less dangerous.

## Restraining of Bulls

- The bull should be ringed by the time of about one year of age, by which time he begins to show his strength.

- A smaller ring can be put at this age, and can be replaced with bigger one when he matures.

- Nose rings are made in two semi-circular pieces hinged together and are of aluminum, copper or some alloy which does not rust.

- The free end of the two parts either, dovetails into one another or are in a form of point and socket, secured either by a flush spring or by a screw with counter sunk head, so that the joint is smooth.

- Since the nose is extremely sensitive to touch, ring in the nose enables the attendant to keep the neck extended and the head raised while restringing or parading.

- Nose ring is an essential item in control of bulls. Bull leading poles can be conveniently hitched to the nose ring and this is mostly felt necessary also.

- The bull can be effectively controlled by means of a chain or rope around the horns threaded through the nose ring.

## Training of Bulls

- The young bulls should be trained for handling and leading.

- It is much easier to maintain control on a mature bull if he was properly trained when young.

- Even when the bull is 4-6 months of old a simple halter may be put over his face and he be accustomed to handling.

- After the nose ring is put he should be led either by chain or pole.

- While leading, the attendant should never walk in front of the bull, but must lead from the side holding the nose always higher than natural level.

- If the nose is allowed to drop, the bull may get inclination to butt.

- While handling and leading, all bulls should be considered as potentially dangerous and no complacency should be shown at any time even in case of old as well acquainted bulls.

## Exercise for Bulls

- Growing as well as mature bulls should be regularly be exercised. So that they do not put on fat and thus remain in thrifty condition.

- These will also helping keeping their toes well worn. Over grown toes may hinder walking as well as mounting behavior of bulls.

## Care of Mature Bulls

- Breeding bulls should never be allowed to run with the herd. They should be housed in separate paddock, individually.

- This helps in controlling number of services by the bulls for recording breeding data.

- The hair around the prepuce should be trimmed periodically.

- The hair should not be clipped too close which may cause irritation and itching to the prepuce. About 1 cm length may be ideal.

## Maintenance of Sexual Libido of Bulls

- There are several factors which can reduce libido in bulls like young or old age, inexperience, tiring exercise, or too frequent usage, semen collection at unusual places in un favourable conditions and using unsuitable fittings, faulty feeding, obesity or run down condition, inherent defects, temporary injury or chronic defect of legs, back and penis. All such problems should be rectified as soon as noticed.

- Some bulls are sensitive to artificial vagina whereas others seem able to withstand considerable rough handling.

- The well known reflexes of mounting the cow, projecting the penis, thrusting and ejaculation can easily be retarded or even inhibited in a bull by unnatural method of handling.

- Majority of the bulls serve well in familiar surrounding and are handled by the same attendant provided these are associated with previous satisfactory experience.

- The sexual reflexes can be inhibited by painful, uncomfortable or even distractive situation.

- In a sensitive bull, inhibition may develop quickly, even when collections are taken carefully.

- The animal should be give rest from collection for as long as possible when inhibition starts developing. This can be overcome by changing the surrounding.

- Overwork is common in young bulls allowed free access to cows and heifers.

- The number of services and not the number of cows served is the important consideration. No bull should be allowed to serve each cow more than twice in a heat period.

- A young bull may be placed with 2 or 3 cow per week and it can be put into service after 2-2 ½ years of age.

- A mature bull may ejaculate many times per week without effect on libido or semen quality.

- The bull with reduced libido should be teased by delaying the service. Bulls become bored in their surrounding, particularly if in small paddock and may lose interest.

- Presence of another bull or change in the surrounding will overcome this problem.

- Summer stress leads to low sexual libido and poor semen quality, especially in purebred exotic and crossbred bulls.

- To overcome such problems during summer, bulls should be housed in cool, well ventilated dry sheds.

- Showering or splashing cold water on bull 2 or 3 times during hot part of the day and protection against direct and reflected radiation were found to be very useful.

## Feeding of Mature Bulls

- A good rule to feed mature bull is to feed daily about 1 kg hay and ½ kg concentrate per 100 kg body weight.

- Thus a 400 kg bull should get 4 kg hay and 2 kg concentrate.

- These amounts should be adjusted according to the body condition of various bulls because there is individual variation in response.

- Excess fatness in mature bull should be avoided at all costs as it reduces libido and may cause severe stress and strain on their feet and legs.

- Excess calcium in bull ration can cause problem particularly in older bulls.

- When legume roughage is fed the concentrate mixture should not contain a calcium supplement.

- Generally bulls do not lose calcium and in time excess calcium may cause vertebra and other bones to fuse together.

- Therefore bulls may need a different concentrate mixture than the milch cows.

# Livestock Health Management

Spring can bring about warm, wet weather. Fluctuating temperatures and wet, muddy conditions can affect the health of your livestock. Livestock may need special care and extra energy to cope with the change in weather. Below are some tips for spring healthcare management:

- Be sure to always provide your animals with access to adequate, dry shelter. Shelter will allow animals to escape from harsh weather conditions.

- Schedule a regular vaccination and deworming program to promote good health in your animals. A veterinarian can assist in helping to develop a vaccination and deworming program for your animals. Spring is the time of the year that worm eggs are hatching and pastures are being infected due to the warm temperatures.

- Make sure to feed livestock a diet that meets the nutritional requirements of your animal to cope with weather conditions.

- Be sure to provide your animal with access to clean water.

Though spring has not fully sprung, it's not too early to consider summer healthcare management as well. Summer brings about hot, humid conditions and some animals" natural ability to cool themselves is limited. Some of the practices for spring healthcare can also be used for summer.

- Providing shelter is essential to keeping animals cool from the hot sun, lowering the risk of heat stress.

- Cool water should be provided at all times to encourage water intake. Pond water may become hot on the surface and unappetizing to drink; a supplemental source is necessary.

- If animals are housed in barns, ventilation is necessary to keep animals from becoming overheated.

No matter what season it is, remember to take care of your livestock as well as yourself, especially during hot summer months, to prevent heat exhaustion or stress. Be sure to consume enough water to stay hydrated and take frequent breaks when working outdoors.

## References

- Livestock-management, livestock-management: agric.wa.gov.au, Retrieved 31 July, 2019

- "Glossary of Meat Terminology". USDA Agricultural Market Service. Accessed. Archived from the original on March 22, 2009. Retrieved May 2, 2007

- Dehorning, knowledge-centre: futurebeef.com.au, Retrieved 1 January, 2019

- Gottardo, Flaviana; et al. (November 2011). "The dehorning of dairy calves: practices and opinions of 639 farmers". Journal of Dairy Science. 94 (11): 5724–5734. Doi:10.3168/jds.2011-4443

- Animhus-cattle-daily%20operation, animal-husbandry: agritech.tnau.ac.in, Retrieved 2 February, 2019

- Ranching, encyclopedia: nationalgeographic.org, Retrieved 3 March, 2019

- O'Barry, Ric (17 April 2014). "JEDOL AND SAMPAL ARE FREE - WE HAVE PROOF!". Dolphin Project. Retrieved 1 June 2014

- Livestock-and-manure-management, activity: ccacoalition.org Retrieved 4 June, 2019

- Larsson, D.G. (2014). "Antibiotics in the environment". Upsala Journal of Medical Sciences. 119 (2): 108–112. Doi:10.3109/03009734.2014.896438. PMC 4034546. PMID 24646081

- Livestock-disease-management, content, climatetechwiki.org, Retrieved 5 July, 2019

- University of Nebraska, Lincoln (October 2015). "Veterinary Feed Directive Questions and Answers". UNL Beef. Retrieved 14 March 2017

- Livestock-health-management-tips: drovers.com, Retrieved 6 August, 2019

# Index